Universal Optimization and its Applications

Alexander Bolonkin

USA, LULU 2017

Title: **Universal Optimization and its Applications**
Author: **Alexander Bolonkin,** abolonkin@gmail.com
ISBN: 978-1-387-22387-9

The book consists of two parts. The first part describes new method of optimization that has the advantages at greater generality and flexibility as well as the ability to solve complex problems which other methods cannot solve. This method, called the "Method of Deformation of Functional (Extreme)", solves for a total minimum and finds a solution set near the optimum. Solutions found by this method can be exact or approximate. Most other methods solve only for a unique local minimum. The ability to create a set of solutions rather than a unique solution has important practical ramifications in many designs, economic and scientific problems because a unique solution usually is difficult to realize in practice.

The second part of the book is devoted to applications of this method to technical problems in aviation, space, aeronautics, control, automation, structural design, and etc.

Copyright@ 2017 by author
Published Lulu in USA: www/lulu.com

Universal Optimization and its Application

(Chapters 1 – 2 are translated from Russian book "New Methods Optimization and its Applications, Moscow, MHTU, 1972)

TABLE OF CONTENTS

About author
Introduction. Abstract

Part one. Mathematical base the methods of optimization   8

Chapter 1. Methods of β- and γ-finctionals   13

1. Method of β-functional
2. The method of combining extremes. Algorithm 3.
3. Note of the γ-functional
4. The application of β-functional in the theory of extrema the functions of a finite number the variables and in optimization problems described by ordinary differential equations
5. β-functional method in the construction of minimizing sequences
    Appendix to Chapter 1
    References to Chapter 1.

Chapter II. Methods of α-functional   41

1. Theory of α-functional. Estimations.
2. The general principle of reciprocity the optimization problems
3. The application of α-functional to known optimization problems
4. The method of reverse lookup
5. The method of combining the extrema in the conditional minimum problems
6. A generalization of Theorem 3.1 to the case of discontinuous $\psi(t, x)$
7. Optimization tasks described ordinary differential equations with restrictions
8. Optimization of discrete systems
9. Optimization of functionals depending on the intermediate values
10. Note on the equivalence of different forms of variational problems
    Appendix to Chapter II.
    References to Chapter II.

Chapter III. MaxMin method (Chapters 3 -11 not included In this book)

1. Principle of the MaxMin method
2. Application of the method of the MaxMin to optimization problems described by ordinary differential equations
3. The method of MaxMin as a method of evaluation of solutions to a system of ordinary differential equations.
4. Application of MaxMin in the study of stability the ordinary differential equations.

5. Application of the MaxMin method to problems distributed parameters and to discrete problems.
   References to Chapter III.

Chapter IV. Numerical implementation of some algorithms α-functional and maximin. Other numerical methods.

1. Numerical implementation maximin method for problems described by ordinary differential equations.
2. The method of steep descent in space of conditions for optimization problems described by ordinary differential equations.
3. Synthesis problems.
4. Construction of the approximate optimal control synthesis.
5. The method of the pieces optimization
6. Some methods for solving boundary problems in the theory of optimal control
7. Descent method along the admissible set in the search for an extremum the functions of a finite number of variables
8. Note on approximate methods of constructing a function $\psi(t, x, y)$.
   References to Chapter IV

Chapter V. Switching Modes

1. Statement of the problem. Basic definitions. Search Methods for minimums.
2. The problem of the most advantageous shape for airbrake.
   References to Chapter V.

Chapter VI. Extremals in optimal control problems.

1. Preliminary remarks.
2. Special extremals.
3. The conversion method in the singular extremals.
4. Sliding modes as a special case of singular extremals.
   Annexes to Chapter VI.
   References to Chapter VI.

Chapter VII. Special extremals and the solvability of boundary value problems in optimal control.

1. Boundary value problems in the theory of optimal control.
2. The existence of special modes - the main reason it is impossible to solve many boundary problems in the framework of the previous methods.
3. Conjugate points - the source of the local "pits" and false solutions
4. Some of the recommendations
   References to Chapter VII.

Part Two. APPLICATION OF METHODS α-, β- functionals and Maximin for technical problems

Chapter VIII. Some tasks of automation

1. The energy minimization problem of signal
2. The problem linear in the phase coordinates and non-linear in controls
3. The problem of the precise regulation.
4. The problem of the minimum fuel consumption.
   References to Chapter VIII.

Chapter IX. Some problems of flight dynamics.

1. The problem of the minimum of the integral heat when the space ship is entering in atmosphere.
2. The challenge of flying at the maximum range missiles (aircraft) with a constant thrust engine.
3. The challenge of flying at maximum range missiles (airship) engine with an adjustable constant power.

Chapter X. Application of α-functional extreme to problems of combinatorial type

1. Statement of the Problem
2. The assignment problem (the problem of choice)
3. The problem of integer programming.
   References to Chapter X

Chapter XI. The problem of counteraction

1. The problem with the opposition (conflicts of players).
2. Numerical methods for finding the solution of game.
3. The methods of synthesis of opposition task.
   References to Chapter XI.

Attachments (Applications): (these chapters are added in this book)

1. Chapter 12. Optimal Thrust Angle of Aircraft. — 73
2. Chapter 13. Optimal Trajectories of Aerospace Vihicles. — 87
3. Chapter 14. Long Distance Bullets and Shells. — 116
4. Chapter 15 . Deep Penetration Bombs; — 138
5. Chapter 16. Design of Optimal Regulators. — 144
6. Chapter 17. Impulse Solutions in Universal Optimization Theory. — 164
   General references — 170

About the Author

Bolonkin, Alexander Alexandrovich (1933-)

Alexander A. Bolonkin was born in the former USSR. He holds doctoral degree in aviation engineering from Moscow Aviation Institute and a post-doctoral degree in aerospace engineering from Leningrad Polytechnic University. He has held the positions of senior engineer in the Antonov Aircraft Design Company and Chairman of the Reliability Department in the Clushko Rocket Design Company. He has also lectured at the Moscow Aviation Universities. Following his arrival in the United States in 1988, he lectured at the New Jersey Institute of Technology and worked as a Senior Researcher at NASA and the US Air Force Research Laboratories.

Bolonkin is the author of more than 250 (2015) scientific articles and books and has 17 inventions to his credit. His most notable books include The Development of Soviet Rocket Engines (Delphic Ass., Inc., Washington , 1991); Non-Rocket Space Launch and Flight (Elsevier, 2006); New Concepts, Ideas, Innovation in Aerospace, Technology and Human Life (NOVA, 2007); Macro-Projects: Environment and Technology (NOVA, 2008); Human Immortality and Electronic Civilization, 3-rd Edition, (Lulu, 2007; Publish America, 2010); Femtotechnologies and Revolutionary Projects, LAMBERT, 2011; Innovations and New Technologies (v.2), Lulu, 2013; Life and Science, LAMBERT, 2011; Small Non-Expensive Electric Cumulative Thermonuclear Reactors, 143 ps., Lulu,2017; Preon Interaction theory and Model of Universe (v2), 103 ps., Lulu, 2017, etc. .

Abstract

The book consists of two parts. The first part describes new method of optimization that has the advantages at greater generality and flexibility as well as the ability to solve complex problems which other methods cannot solve.

This method, called the "Method of Deformation of Functional (Extreme)", solves for a total minimum and finds a solution set near the optimum. Solutions found by this method can be exact or approximate. Most other methods solve only for a unique local minimum. The ability to create a set of solutions rather than a unique solution has important practical ramifications in many designs, economic and scientific problems because a unique solution usually is difficult to realize in practice.

This method has the additional virtue of a simple proof, one that is useful for studying other methods of optimization, since most other methods can be delivered from the Method of Deformation.

The mathematical methods used in the book allow calculating special slipping and breaking optimal curves, which are often encountered in problems of optimal control. The author also describes the solution of boundary problems in optimization theory.

The mathematical theory is illustrated by several examples. The book is replete with exercises and can be used as a text-book for graduate courses. In fact the author has lectured on this theory using this book for graduate and post-graduate students in Moscow Technical University named Bauman.

The second part of the book is devoted to applications of this method to technical problems in aviation, space, aeronautics, control, automation, structural design, economic, games, theory of counter strategy and etc. Some of the aviation, aeronautic, and control problems are examined: minimization of energy, exact control, fuel consumption, heating of re-entry space ship in the atmosphere of planets, the problems of a range of aircraft, rockets, dirigibles, and etc.

Some of the economic problems are considered, for example, the problems of a highest productivity, the problem of integer programming and the problem of linear programming.

Many economic problems may be solved by the application of the Method to the Problems of non-cooperative games. In given book is only translated Chapter 1 – 2 from initial author Russian book "New Methods of Optimization and its Application", Moscow, MVTU named Bauman, 1972, 220 ps.

The new third part of the book contains solutions (1988-2016) of complex problems: optimal thrust angle for different flight regimes, optimal trajectories of aircraft, aerospace vehicles, and space ships, design of optimal regulator, linear problems of optimal control.

This book is intended for designers, engineers, researchers, as well as specialists working on problems of optimal control, planning, or the choosing of optimal strategy.

For engineers the book provides methods of computation of the optimal construction and control mechanisms, and optimal flight trajectories.

In addition, the book will be useful to students of mathematics, general engineering, and economic. additional sets is proposed.

The method (a) reduces the initial complex problem of optimization to series of simplified problems, (b) finds the subsets containing the point of global minimum and finds the subsets containing better solutions that the given one, and (c) obtains a lower estimation of the global minimum.

Part 1

Mathematical Base of the Universal Optimization

Introduction

The classical approaches this problem is following:

Problem A. *Find a minimum of the given function.*

Together with problem A the following problems are considered:

Problem B. *Find a smaller subset contains the all points of the global minimum.*

Problem C. *Find a subset of better solutions where the function is less that given value.*

Problem D. *Find a lower estimation of function.*

These non-classical approach B, C, and D require innovative methods, different from the well-known methods.

The author offers a new mathematical methods for the solution of these problems.

The new methods have turned out to be much more general, so that while solving one of the above problems, another may be solved in passing, which may help in the solution of the former. Thus, if a satisfactory lower estimate found, it can be compared with various engineering solutions and give rise to one very close to the optimum.

This method is applied to many mathematical problems of optimization. For example, functions of several variable, constrained optimization, linear and nonlinear programming, multivariable nonlinear problems described by regular differential equations and equations in partial derivatives, etc.

One can easy get from the given method to many well-known methods of optimization, for example, Lagrangian multiplier method, the penalty function method, the classical variational method, Pontragin's principle of maximum, dynamic programming and others.

At present, the most of researchers in optimization fields are using the traditional optimization problem – find a minimum of the given functional (Problem A). They look a single, local minimum. An engineer, however, is usually interested in a subset of quasi-optimal solutions. He must make sure that the optimum does not exceed a given value (Problem C). Also, a good estimation from below will indicate how far a given solution is from the optimum solution (Problem D). An addition an engineer usually has other considerations that cannot be introduced into a mathematical model or can lead to impractical complications. Approach C provides him with some choice.

Problem D is also of particular interest. If an estimate from bottom closes to the exact infinum of the function is found, the optimization can frequently be reduced to finding a quasi-optimal solution by trial and error.

Solution of the Problem B can significantly simplify the solution of any of the above problems, since it narrows the set containing optimal solution.

These non-classical Problems B, C, and D require innovative methods, different from the well-known method of variational calculus, maximum principle and dynamic programming. This new method is general, so that while solving one of the above problems, another may be solved in passing, which may help in the solution of the former. Thus, if a satisfactory estimate from below has been found, it can be compared with various engineering solutions and give rise to one very close to the optimum.

Our reasoning in this book is not complex. But we are using symbolic of set Theory, which many engineers forget. That way we are given these information in Appendix A of the book.

In Book we are using the double numbering of formulae, theorems and drawings. The first figure in nubbering formule or theorem notes the number of paragraph; the second figure is number formula or theorem in this paragraph. The first figure of drawings means the number of chapter, the second is the number of drawing.

Some information from theory of sets.

The concept of **set** (family) is one of the primary in science and is not defined through other simpler concepts. This is the unification of objects on some basis. Examples of set: a lot of pages of the book, many stars and planets, many students, many rational numbers, etc.

Sets are usually denoted by uppercase letters: X, Y, M, N, P. Objects composing a set are called by its elements denoted by lowercase letters x, y, ... The sign $\in$ denotes belonging $x \in X$. It is read: the **element** x belongs to the set X. If the element x not belong Y, then one write $x \notin Y$. This is reading: the element x not belong for set Y. If set X contains only elements x we write $X = \{x\}$. If the set has the limited number of elements, we spoke: one is limited, in other case the set is infinity.

If the set do not has any or single element, we named it by **empty** and denoted $\varnothing$. If all element from set A belong the set B, we speak: the set A include into B, or set B contains the set A, or set A is part (subset) of set B. We write $A \subset B$ or $B \supset A$. If it is possible the case A = B, we write $A \subseteq B$ or $B \supseteq A$.

The set may be denote the next:
a) Enumeration of all its elements. For example: set {0, 1, ... ,9}.
b) The description of the limited property. Example: $M = \{x : \beta(x) \geq \beta(\bar{x})\}$ - this is set from elements, which are satisfying the inequality $\beta(x) \geq \beta(\bar{x})$, where $\bar{x}$ is given element, β(x) is some real function.
The set may be selected more complex way. For example, the set may depends from variale y ∈ Y. Example: $M = \{x : \beta(x, y) \geq \beta(\bar{x}(y), y), y \in Y\}$;
b) The set may be selected by operations over sets. Examples:
Summiring or acssociation of two sets A and B. Denote A + B or $A \cup B$. This set contains all elements belong to sets A and B (Fig. !).

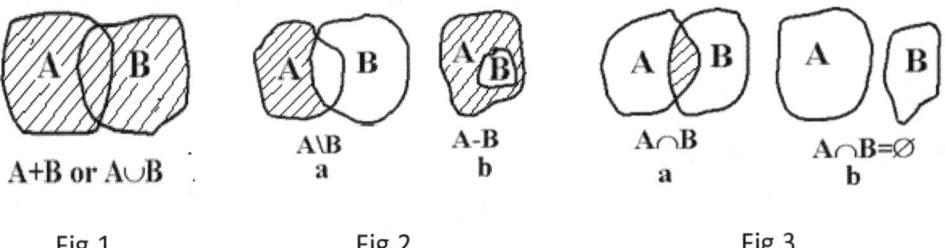

Fig.1 Fig.2 Fig.3

Difference of two sets A and B is named the set which contains all elements of set A not include in set B and not contein any others elements (Fig.2a). The difference sets is noted A − B or A\B. The first designation is used only $A \supseteq B$ (Fig.2b).

Intersection of sets *A* and *B* is called the set conteins simultaneously from the elements of *A* and *B*. Their designation is $A \mid B$ or $A \cdot B$. If $A = A_1 A_2 ... A_n$, then $A = \bigcup_{i=1}^{i=n} A_i$. If $A \mid B = \emptyset$, then sets *A*, *B* named **nonintersection** (Fig.3b).

Assume there is two sets $A = \{a\}$ and $B = \{b\}$. The set of the ordered pair of elements (a,b) where *a* belong *A* and *b* belong *B* are calling Decard's **product** of sets *A* and *B* and designation $A \times B$.

Assume $c = (a,b)$ is the element of sets $A \times B$. The element *a* is **projection** of element *c* on set *A*. If $E = A \times B$ then projection *E* on *A* is named the set of elements from *A*, which are projections of elements from *E* on *A*. Designation is $pr_A E$.

Cross-section $x = a$ set *E* is named the set of elements $y \in B$, which $(a,y) \in E$.
Real numer *b* is named **majorant** (respectively **minorant**) the set *A* of the real numbers, if $a \leq b$ (respectively $b \leq a$) for any $a \in A$. Expression "any" offen definity the simbol $\forall$. The set $A \in R$ (*R* is numerical axis) is named **majorized** if it is limited over and minoranted if it limited below. *A* is not empty. The set is bounded from above and from below by the name of **bounded**.

If the majorant of *A* belong the *A*, it is named **maximum** if set *A* and nominity $\max_{x \in X} A$ or $\max_X A$.

Appropriately for minorant we write **minimum**: $\min_{x \in X} A$ or $\min_X A$.

If sets majorants (minorants) has maximum (minimum), this element is named top (low) **face** of set and nominity:

$$\sup A, \sup_{x \in A} A(x), \sup_X A(x), \inf A, \inf_{x \in X} A(x), \inf_X A(x).$$

Signs *sup, inf* are reading as: *supremum, infinum*. Sometimes it is used the designation:
$$\sup A(x), x \in X \text{ or } \inf A(x), x \in X.$$

The top and low face of set *A* may be not belong the set *A*.

Sometimes people undestend *max (min)* as *local maximum (minimum)*, but *sip (inf)* as *global (absolut) maximum and minimum)*. Maximum $f(\bar{x})$ (here $\bar{x}$ is point of extremum: *point of maximum or minimum*) of real function *f(x)* given on set *X*, which has distance between elements, is named **local maximum**, if it exists the environs of element (point) $\bar{x}$ in which $f(\bar{x}) > f(x)$ for any $x \neq \bar{x}$. Similarly, the local minimum of the function.

The function is defined on an arbitrary set whose values are real numbers is called the **functional**.
It is remaind some properties of operations *sup inf* and *inf sup*. Assume *f(x,y)* is real functin of two variables defined for $x \in A$ and $y \in B$. If $\sup_{x \in A} \inf_{y \in B} f(x,y)$ and $\inf_{y \in B} \sup_{x \in A} f(x,y)$ are exist, then

$$\sup_{x \in A} \inf_{y \in B} f(x,y) \leq \inf_{y \in B} \sup_{x \in A} f(x,y).$$

Assume *f(x,y)* is real function defined on $A \times B$. Point (x_0, y_0), where $x_0 \in A$, $y_0 \in B$ is named the *saddle point*, it has the conditions:
1) $f(x, y_0) \leq f(x_0, y_0)$ for $\forall x \in A$;
2) $f(x_0, y_0) \leq f(x_0, y)$ for $\forall y \in B$.

Thus, for a saddle point $f(x, y_0) \leq f(x_0, y_0) \leq f(x_0, y)$.

Assume *f(x,y)* is functional defined on $A \times B$ and exist $\max_A \min_B f(x,y)$ and $\min_B \max_A f(x,y)$, then nessusary and suffiiently conditins the equal $\max_A \min_B f(x,y) = \min_B \max_A f(x,y)$ consists in next: the function *f(x,y)* must has the **saddle** point. Moreover, if $f(x_0, y_0)$ is a saddle point of functional *f(x,y)*, then

$$f(x_0, y_0) = \max_A \min_B f(x,y) = \min_B \max_A f(x,y).$$

Similarly with local maximum and minimum we can define the local saddle point, when it exists the environs

$A_1 \otimes B_1 \subset A \otimes B$ of point (x_o, y_o) in which $f(x, y_o) \leq f(x_o, y_o)$ for $\forall x \in A_1$ and $f(x_o, y_o) \leq f(x_o, y)$ for $\forall y \in B_1$.

Note: For us more camfotable use other define the saddle point then in theory of games. We define the saddle point as the point has conditions: 1) $f(x, y_o) \geq f(x_o, y_o)$ for $\forall x \in A$; 2) $f(x_o, y_o) \geq f(x_o, y)$ for $\forall y \in B$. The saddle point in our understanding is $f(x_o, y) \leq f(x_o, y_o) \leq f(x, y_o)$ and
$$f(x_0, y_0) = \max_B \min_A f(x, y) = \min_A \max_B f(x, y).$$ Here arguments of function are understanding.

Properties of inequalities.

1. Adding (substracting) to both sides of inequality the same numbers (constant c). If $a > b$, then $a + c > b + c$.
2. **A**dding inequalities of the same meaning. If $a > b$, $c > d$, Then $a + c > d + d$.
3. Substrating the inequalities the opposet meaning. If $a > b$, $c < d$, then $a - c > b - d$.
4. Multiplication of the inequility in constant c. If $a > b$ and $c > 0$, then $ac > bc$. If $c < 0$, then $ac < bc$.
5. Multiplication of the inequility in same meaning. If a, b, c, d are positive and $a > b$, $c > d$, then $ac > bd$.
6. If $a > b > 0$, then for any digital n $a^n > b^n$.
7. If $a > b > 0$, for any digital $n \geq 2$, $\sqrt[n]{a} > \sqrt[n]{b}$.
8. inversion of inequalities. If $a > b$ and $a \neq 0$, $b \neq 0$, then $1/a < 1/b$.

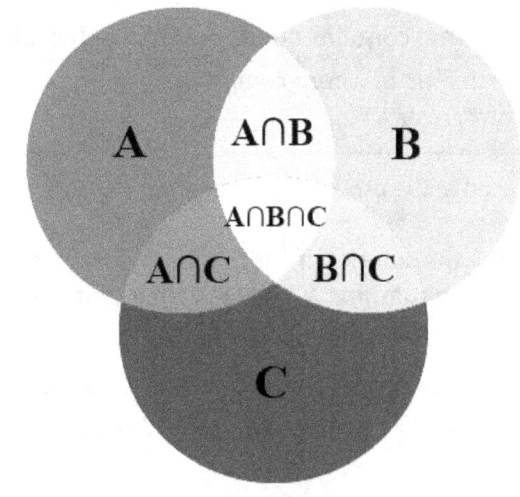

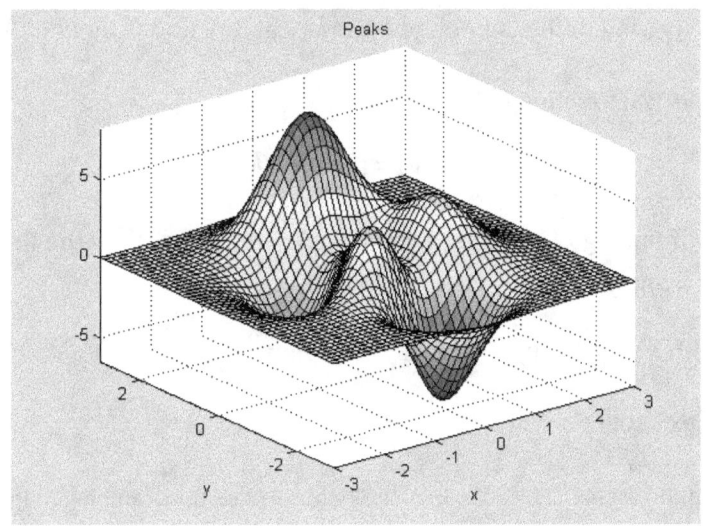

Chapter 1

Methods of β and γ functions

§1. Methods of β – functions

1. Statement of the Problem. Main theorems. Algorithm 1.

1⁰. Statement of the Task. Assume that the state of the system is described by element *x*. A series of these elements form the set *X={x}*. The numerical function *f(x)* (or *I(x)* - functional) is defined and bounded by its lower estimate over *X*. The relationships and limitations imposed on the system yield a subset $X^* \subseteq X$.

Traditionally the problem of optimization has been set as follows:

A. Find a point of the global minimum[*] *x** of the function *f(x)* over the set *X**.

We shall also consider the following problems:

B. Find a smaller subset $M \subset X^*$ that contains point *x** of global (absolute) minimum, $x^* \in M$.
C. Find a subset $N \subset X^*$ on which *f(x)≤ c*, where *c ≥ f(x*)*.
D. Find the lower estimates of *f(x*)* over *X**.

---
[*] Problem of maximum is reduced to the problem of minimum by change the siat at *f(x)*.

We will name the point (element) *x* the solution if *x* is result any presses, procedure, calculation, or reasoning. It not means that *x* is point of optimum. We will tell the point x_1 is better solution than the point x_2, if $I(x_1) < I(x_2)$ and the point of the same solution, if $I(x_1)=I(x_2)$.

For simplicity we assume that the point of the global minimum *x** is single and exists in *X**, but this is not impotent limitation. The most results can be obtained without this assumption. But we assume that $\inf_{X^*} I(x)$ is exist.

Let us introduce a set *Y={y}* and define a bounded numerical function (functional) *β(x,y)* over *X×Y*. We shall call it **β-functional**.[*]
Then we can create a numerical function (a functional)

$$J(x, y) = I(x) + \beta(x, y).$$

Call our initial problem of finding *x** and $I(x^*) = \inf I(x) = m, \quad x \in X^*$ the **Problem 1** and the problem of finding $\bar{x}$ and

$$\bar{J}(\bar{x}(y), y) = \inf[I(x) + \beta(x, y)], \quad x \in X \quad \textbf{Problem 2}$$

We assume that $\bar{x}(y)$ exist over *X×Y*.

---
[*] The feasibility of introducing the set *Y* will be seen in next Chapters (see, for example, Chapter 3).

We deformed arbitrarily our functional *I(x)* by adding *β(x,y)*. Moreover we widened the domain of the deformed functional and arbitrarily defined it on the set *Y*. we should do so in such a way that problem 2 will be easier to solve.

It might seem that this makes no sense because we must find the points of minimum of our initial functional *I(x)*, i.e., solve Problem 1. But it appears that from the solution of the simpler Problem 2 we can obtain information about Problem 1. We can use freedom in choice of the functional *β(x,y)* and the set *Y* for such a deformation of functional *J(x,y)* and the set *Y* that we solve the initial Problem 1, but in an easier way.

2^0. The Fundamental Theorem. The following main theorem establishes the relationship between Problem 1 and 2, as well as between Problems A, B, and C (The Principle 1 of Optimum).

Theorem 1.1. *Distinguishing between the sets containing: (1) The global minimum points, (2) only better solutions than the one given, (3) only worse solutions than one given.*

Assume $X^* \equiv X$, $\bar{x}(y)$ are the points of global minimum in Problem 2. Then:

(1) The points of global minimum in Problem 1 are contained in the set
$$M = \{x : \beta(x,y) \geq \beta(\bar{x}(y), y),\ y \in Y\};$$

(2) The set
$$N = \{x : J + I \leq \bar{J} + \bar{I},\ y \in Y\};$$

contains the same or better solutions (that is over N, we have $I(x) \leq I(\bar{x})$);

(3) The set
$$P = \{x : \beta(x,y) \leq \beta(\bar{x}(y), y),\ y \in Y\};$$

contains the same or worse solutions (that is over P $I(x) \geq I(\bar{x})$).

. Proof of Theorem 1.1.

Statement 3. By subtracting the inequality $\beta(x,y) \leq \beta(\bar{x}(y), y)$ from $I(x) + \beta(x,y) \geq I(\bar{x}(y)) + \beta(\bar{x}(y), y)$ we get $I(x) \geq I(\bar{x})$ over P. Statement 3 of the Theorem 1.1 is proved.

Statement 1 of the Theorem 1.1 is obvious because *X=M+P* and $I(x) \geq I(\bar{x})$ over P, we have $x^* \in M$.
Statement 1 of Theorem 1.1 is proved.

Statement 2. By subtracting the inequality $J \geq \bar{J}$ from $J + I \leq \bar{J} + \bar{I}$ we get $I(x) \leq I(\bar{x})$ over N.
Point 2 of the theorem is proved.

Theorem 1.1 is proved.

If in sets *N* and *P* we write the strong inequality $\beta > \bar{\beta}$, then the set *N* will contain only better solutions and the set *P* will contain worse solutions that $I(\bar{x})$.

Theorem 1.1 is correct when *X*≠X*, but *M, N, P* contain elements from *X**.

Let us focus our attention on the fact that after solving the simpler Problem 2, we distinguished in our set *X* three subset: *M*, which contains a point of global minimum, subset *P*, containing the same or worse solutions, and subset *N*, which contains the same or better solutions.

Consequences:
1. Element $\bar{x}$ is the point of global minimum of the functional over the set *P⊂X*.
2. $\bar{x}$ is the element which gives the maximum of the functional *I(x)* over the set *N⊂X*.
3. If *X*⊆P*, then $\bar{x}$ is the point if global minimum Problem 1 over set *X**. In this case we have *M = {x}*.
4. If *β=β(x), x ∈X*, then

$$M = \{x : \beta(x) \geq \beta(\bar{x})\}, \quad P = \{x : \beta(x) \leq \beta(\bar{x})\}, \quad N = \{x : J + I \geq \bar{J} + \bar{I}\}.$$

Theorem 1 is correct when $X^* \neq X$, but M, N, P contain element from X^*.

5. Let $X^* \neq X$. If $X^* \cap M = \varnothing$, then $I(\bar{x})$ is the lower estimation $I(x)$ over the set X^* (because in this case we have $X^* \subseteq P$).
6. Let $X^* \neq X$. If $X^* \subset N$, then $I(\bar{x})$ is the top estimation $I(x) \leq I(\bar{x})$ over the set X^*.

If $\bar{x} \in X^*$, the sets M, N, P will always contain at least one element from the set X^*. This element is $\bar{x}$.

Remarks:

1. $N \subseteq M$. The proof: Let us denote $\overset{o}{P} = P - \{\bar{x}\}$. Then $\overset{o}{P} \cap N = \varnothing$, because over $\overset{o}{P}$ we have $I(x) > I(\bar{x})$ and over N we have $I(x) \leq I(\bar{x})$. But $N \subset X$ and $M = X - \overset{o}{P}$. Hence $N \subseteq M$.
2. Assume the definitions of the sets N, P (see Theorem 1) contain strong inequalities. Then the set N will contain better solutions then $\bar{x}$ and the set P contain only worse solutions, compared to $\bar{x}$.
3. We can use the dependence of the sets M, N, P from y in order to change the "dimensions" of these sets.
4. *β* - functions exist and their number is infinite.
   The last statement is obvious because we can define *β*-functionals over the set $X \times Y$ in any possible way.

The theorem 1 gives the Algorithm 1 (a *β*-functional method for finding the subsets that contains the points of global minimum or better solutions).

Algorithm 1. *Define $\beta_i(x,y)$ so that Problem 2 becomes easier to solve, and find sets M_i and N_i. Then $M = \cap M_i$ (that is not empty) is the set that contains the points of global minimum and $N = \cap N_i$ (if that is not empty) is subset contains $\min \{I(\bar{x}_i)\}$ or better solutions.*

Note: The getting M is more "narrow" (contains less points x) subset then initial M. That means the finding x^* is easier. The decreasing of M is especially important in a "method of dynamic programming" because it is decreasing the number of computation.

Theorem 1.2. (The lower estimate) *Let us assume that β(x,y) is a defined and bounded functional over $X \times Y$ then the lower estimate over X is*

$$I(x) \geq [I(\bar{x}(y)) + \beta(\bar{x}(y), y) - \sup_X \beta(x, y)] \quad \text{for} \quad \forall y \in Y. \tag{1.1}$$

Proof of Theorem 1.2. By adding the inequality $I(x) + \beta(x, y) \geq I(\bar{x}(y)) + \beta(\bar{x}(y), y)$ and $-\beta(x, y) \geq -\sup_X \beta(x, y)$ over X, we get the estimate (1.2).

Remarks:

1. For case $\beta = \beta(x)$ the estimate (1.1) is

$$I(x) \geq \inf_X J(x) - \sup_X \beta(x), \tag{1.1'}$$

2. When $X \neq X^*$ the estimation (1.1) is correct over X^*, because $X^* \subseteq X$. In this case we can use the better estimations:

$$I(x) \geq \inf_{X^*} J(x) - \sup_X \beta(x), \quad I(x) \geq \inf_X J(x) - \sup_{X^*} \beta(x), \quad I(x) \geq \inf_{X^*} J(x) - \sup_{X^*} \beta(x), \tag{1.1''}$$

When we found the set M for B_i the following estimate may be used
$$I(x) \geq \inf_{X^*} J(x) - \sup_{M} \beta(x), \qquad (1.1''')$$
The proof of (1.1'), (1.1''), (1.1''') is same the proof of theorem 1.2.

3. Dependence of the estimate (1.1) from y may be used for its improving
$$I(x) \geq \sup_{y}[\inf_{x^*} J(x) - \sup_{x} \beta(x)], \qquad (1.1^{IV})$$

When we use the estimates (1.1') - (1.1^{IV}) we decide the problem $\grave{\beta} = \sup_{X} \beta$. It may be used for finding sets M, M, P, by theorem:

Theorem 1.3. Assume X=X*, $\bar{x}$ is point of a global minimum in the problem $\grave{\beta} = \sup_{X} \beta$,

Then:

1) The points of global minimum in Problem 1 are contained in the set
$$M(y) = \{x : I + \beta \leq \grave{I} + \grave{\beta}, \quad y \in Y\}$$

2) The set
$$N(y) = \{x : \beta - I \geq \grave{\beta} - \grave{I}, \quad y \in Y\}$$
contains the same or better solutions.

3) The set
$$P(y) = \{x : I + \beta \geq \grave{I} + \grave{\beta}, \quad y \in Y\}$$
contains the same or worse solutions.
Here is $\grave{I} = I(\grave{x})$.

Proof of Theorem 1.3.
Statements 1, 3. By subtracting the inequality $\beta \leq \hat{\beta}$ from $I + \beta \geq \hat{I} + \hat{\beta}$ we get $I \geq \hat{I}$ over set P.
Statement 1 follow from this.
Statement 2. By subtracting the inequality $\beta \geq \hat{\beta}$ from $\beta - I \geq \hat{\beta} - \hat{I}$ and multiply this result by -1, we get $I \leq \hat{I}$ over N. The theorem 1.3 is proved.

Remark:
For proof of the theorems 1.1-1.3 the existence of x, $\bar{x}$, $\grave{x}$ is not important, but corresponding *inf* and *sup* must be existed.

Example 1.1.

Find minimum of functional
$$I = -e^{-x^4}\cos x^2 - \frac{0.1}{x^2 - 0.2x + 1}, \quad -\infty < x < \infty, \qquad (1.2)$$

Solution. Take

$$\beta(x) = \frac{0.1}{x^2 - 0.2x + 1}.$$

Then

$$J = I + \beta = e^{-x^4} \cos x^2.$$

The minimum of this J is obvious: $\bar{x} = 0$.

Consequently from theorem 1.1 we got the point of the global minimum is in set

$$M = \{x: \beta(x) \geq \beta(0)\} \quad \text{or} \quad \frac{0.1}{x^2 - 0.2x + 1} \geq 0.1.$$

The solution of this inequality is $0 \leq x \leq 0.2$. It's not difficult to find the point of global minimum in this small interval by any known method.

We get the lower estimate (theorem 1.2) $J(0) - \sup_x \beta = -1 - 0.101 = -1.101$.

Value $I(0) = -1.100$. We see $I(x)$ for $x = 0$ is very close to global minimum.

Example 1.2

Find minimum

$$I = -\frac{0.1}{x^2 - 2x + 10} + \cos 4\pi x - 4 \cos 2\pi x, \quad -\infty < x < \infty \quad (1.3)$$

Solution: We take $\beta(x) = -\cos 4\pi x + 4 \cos 2\pi x$. Then

$$J = I + \beta = \frac{0.1}{x^2 - 2x + 10}, \quad \bar{x} = 1.$$

This solution is global minimum of Problem 1 over set $P = \{x: \beta(x) \leq \beta(1)\}$ or $-\cos 4\pi x + 4 \cos 2\pi x \leq 3$.

We transform this inequality in*[)] (*[)] See "Handbook of Mathematics" by K.A. Bronshtein, Moscow, 1956, p.184 (Russian)).---$8\sin^4 \pi x \leq 0$.

We see $P = \{x: |x| < \infty\}$. Therefore $P = X^*$. That means (see Consequence 1) $\bar{x} = 1$ is point (and alone) of global minimum of the functional (1.3).

Example 1.3.

More full, we are demonstrating the new method on following simple functional.

Find the absolute minimum of the functional

$$I = 2x^4 + x^2 - 2x + 1 \quad \text{on the set } X^* = \{x: |x| < \infty\}. \quad (1.4)$$

It is a simple example, which can be solved using well-known methods. For example, take the first derivative, make it equal to zero. Solve an algebraic 3-d order equation (it may not be a simple task) and then analyze the points so found with respect to maximum and minimum.

We shall try to solve this example by the above method as it follows from algorithm 1.

Let us introduce a series $\beta_i(x)$. As follows from Theorem 1.1 we have the sets M_i:

1) Take $β_1=2x$. Then
$$J = I + β_1 = 2x^4 + x^2 + 1, \quad \bar{x} = 0, \quad \text{from} \quad β \geq \overline{β} \quad \text{we have} \quad M_1 = \{x : x \geq 0\} \ .$$
ψ_{x_i} As we see the domain which contain a global minimum have become less in two times.

2) Take $β_2 = -x^2 + 2x$. Then
$$J = I + β_2 = 2x^4 + 1, \quad \bar{x} = 0, \quad \text{from} \quad β \geq \overline{β} \quad \text{we have} \quad M_1 = \{x : 0 \leq x \leq 2\} \ .$$
Our interval contained a global minimum is only 0≤x≤2. For given $β_2$ we can use an estimation of the functional which follows from Theorem 1.2.

$$I(x) \geq J(\bar{x}) - \sup_X β_2(x) = 1 - \sup_X(-x^2 + 2x) = 1 - 1 = 0,$$

where the point of supreme of $β$ is $\acute{x} = 1$.
From theorem 1.3 we have the additional set M: $M_3 = \{x : J(x) \leq J(\acute{x})\}$ or $M_3 = \{x : |x| \leq 1\}$.
As we see the set $M = M_2 \cap M_3 = \{x : 0 \leq x \leq 1\}$, The global minimum of this problem is in the interval 0≤x≤1.

3) Take $β_3 = 2x^2 + 2x - 0.5$. Then $J = I + β_3 = 2x^4 - x^2 - 0.5$. From inf J we have $\bar{x}_{1,2} = \mp 0.5$.

4) Find for point x_1 set M:
$$\bar{x}_1 = -0.5, \quad M_4 = \{x : -0.5 \leq x \leq 1.5\},$$

$$\bar{x}_2 = 0.5, \quad M_5 = \{x : 0.5 \leq x \leq 0.5\} \ .$$

The estimation gives $I(x) \geq 3/8 - 0 = 3/8$.
We see that the diameter of the set $M = \cap M_i$ decreases until reduces in the point $\bar{x} = 0.5$. Therefore this point is one of the absolute minimum of the Problem 1 and $I(0.5) = 3/8$.

The geometric illustration of Theorem 1.1 is given in fig, 1.1 for single variable. The curves $I(x)$, $J(x)$, $β(x)$, $I(x)+0.5\, β(x)$ and point $\bar{x}$ are drawn. There are the sets M, N, P. P is set x, where $β(x) \leq β(\bar{x})$, M is set $X \backslash P$ and N is set x, where $J(x) + 0.5β(x) \leq J(\bar{x}) + 0.5β(\bar{x})$.
We can see that $N \subset M$.
In fig.1, 2 we see sets M, N, P for the case when $I(x_1,x_2)$ is function of two variables x_1 and x_2.

2. About Convergence of Algorithm 1.

Consider condition of convergence $\inf\limits_{x \in X} J(x), \ x \in X$ to $\inf\limits_{x \in X^8} J(x), \ x \in X^*$ and $\bar{x}$ to x^* for Algorithm 1. When we have the succession $β_i(x)$, $I = 1,2,...$ This succession gives the succession of the sets M_i, N_i and values of functionals $J(\bar{x}_i)$.

The succession $\{\inf J(\bar{x}_i)\}$ for $i \to \infty$ is monotonous decreasing and bounded of bottom, that's way it has a limit. If this limit equals one of lower estimates, that $J(\bar{x}) = I(x^*)$.
Let us to consider now convergence of diameter $d(M)$, $d(N)$ of sets $M = \cap M_i$, $N = \cap N_i$ for $I \to \infty$.
This convergence is also monotonous decreasing and bounded of bottom: $d \geq 0$. Therefore it has a limit.
We have got the following simply criterion of convergence

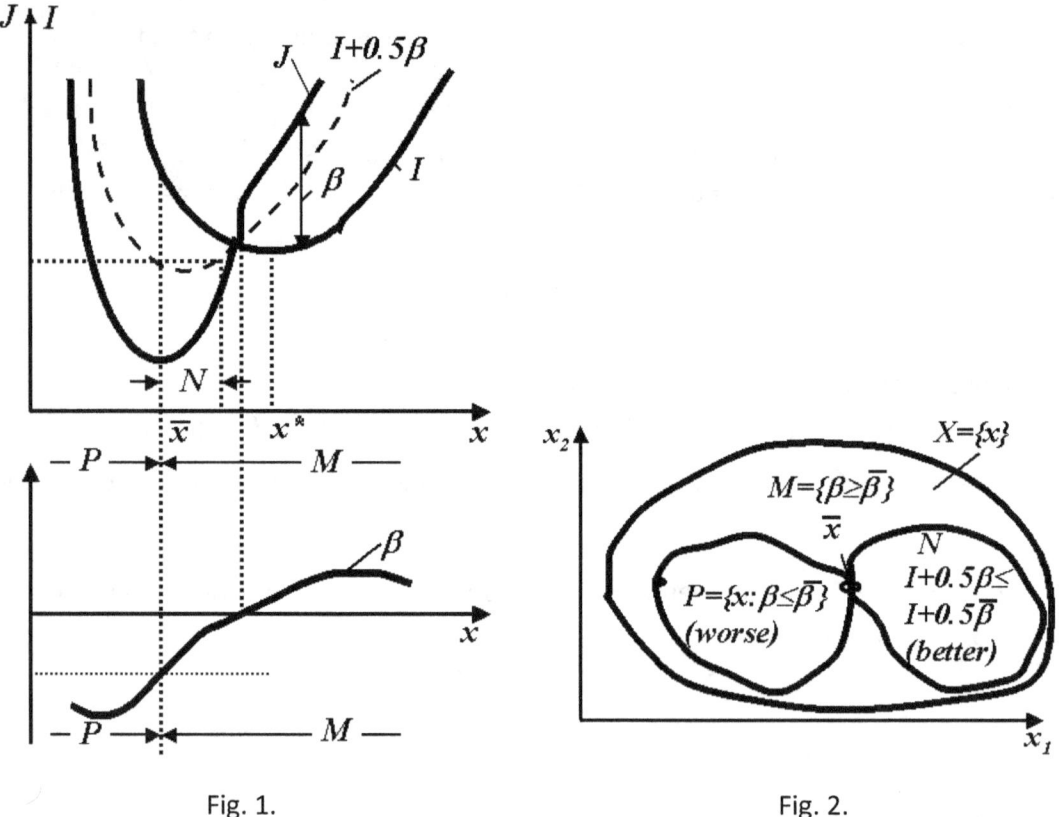

Fig. 1.   Fig. 2.

Fig.1.1. Geometric illustration of Theorem 1.1 for case of single variable.
Fig.1.2. Sets M, N, P for case of two variable.

Theorem 1.4. Assume, the point of the absolute minimum of functional I(x) over set X=X* is single.
If $d(M) \to 0$, than $x = \lim M(i) = x^*$, $l \to \infty$.

In this case the set contained of point of global minimum $M = \cap M_i$ decrease in point. Therefore, this point is the point of the absolute minimum of Problem 1.

Let us take succession of function $W_s(x)$, $s = 1,2,...$. Take $\beta_i(x)$ as

$$\beta_i = \sum_{s=1}^{i} c_s W_s(x) \qquad (1.5)$$

where c_s is constants.
We will take these constants c_s from condition

$$\Delta_i = \min_c [I(\bar{x}_i) - \inf_X J_i(x) + \sup_X \beta_i(x)].$$

The value Δ_i is difference functional from its lower estimate. Other words value Δ show how much value $I(\bar{x}_i)$ differs from optimum. We name this number Δ-estimate (*delta-estimate*). It is obvious that succession $\{\Delta_i\}$ is monotonous decreasing because every next sum (1.5) contains previous sum. It is also limited of bottom ($\Delta_i \geq 0$). Therefore the succession $\{\Delta_i\}$ converge.

From destination Δ_i we get the following

Theorem 1.5. If $\Delta_i \to 0$ then $\inf\limits_{X} J(x) \to \inf\limits_{X^*} I(x)$.

Theorem 1.6. Assume $X = X^*$, $B_i = c_i B(x)$, $I(x)$, $B(x)$ is continuous and $B(x)$ is limited on X.

Then, if $c_i \to 0$ we have $J(x) \to m = \inf I(x)$ over X^*.

Statement of Theorem 1.6 follows from continuous $J(x)$.

This theorem may be useful for finding of the local minimum of $I(x)$ by way of methods of successive approximations. Assume $c_1 = 1$ and problem $\inf J(x)$ can decided simply. Because functional $J(x)$ is continuous, we can wait, that small change of c give small changing (moving) $\bar{x}$.

Therefore $\bar{x}$ is good the initial approximation for $c_2 < c_1$. It is known, that a good initial approximation is very important for speed of convergence. We come to x^* by decreasing c to 0.

These criterions of convergence may be used for solutions Problem A, B, C, D (see §1,A).

1. Modification of the Theorem 1.1

Over we have considered the case, when we are looking for the additional function $B(x,y)$ such us the problem 2 became simpler for solution. But sometimes it's more comfatable to take such function $J(x,y)$ that the problem $\inf\limits_{X} J(x, y)$ became easy for solution. In this case Theorem 1.1. better to write as following:

Theorem 1.1'. Assume $X^* \equiv X$, $x(y)$ is the point of global minimum in Problem 2 . $\bar{J} = \inf\limits_{X} J(x.y)$

Then

1) The points of global minimum in Problem 1 are contained in the set

$$M(y) = \{x: J - I \geq \bar{J} - \bar{I}, \quad y \in Y\}.$$

2) The set

$$N(y) = \{x: J + I \leq \bar{J} + \bar{I}, \quad y \in Y\}$$

Contains the better or same solutions.

3) The set

$$P(y) = \{x: J - I \leq \bar{J} - \bar{I}, \quad y \in Y\}$$

Contains worse or same solutions.

This Theorem 1.1 is correct if $J = kJ_1$, where k = const > 0.

2. Method of big steps in set of better solutions. Algorithm 2.

From the Theorem 1.1 we can get the following

Algorithm 2 (Method of big steps in set of better solutions)

Take any point x_1 from X^* and create such function $J_1(x)$ that point x_1 is its minimum. Find the set N_1 of better solutions. Take from this set a point x_2 and such function $J_2(x)$ that x_2 is its minimum. Find the set N_2 and so on.

It is obvious that $N_1 \supseteq N_2 \supseteq N_3 \supseteq \ldots$. Let us suppose that result of this process is following - set N_i become point x_N.

Theorem 1.7. Assume X^* is open set, $I(x)$, $J_i(x)$ are continuously and differential (of Freshe) on X^*. Then point x_N is a stationary point of the function $I(x)$ over X^*.

Proof of Theorem 1.7. Assume x_N is point of the minimum of the objective function $J(x)$. Therefore $J'(x_N) = 0$ because $J(x)$ is continuously and differential, x_N is single point N_i on set X^* since this is (see Theorem 1.1')

$$I(x) + J(x) \geq I(x_N) + J(x_N).$$

This means that $J_i(x_N) = \inf_X [I(x) + J(x)]$. The function $I(x)$, $J(x)$ are continuously and differential, hence $I'(x_N) + J'(x_N) = 0$. But $J'(x_N) = 0$, therefore $I'(x_N) = 0$. Theorem 1.7 is proved.

Theorem 1.8. If in point x_N we have

$$\beta(x_N) - I(x_N) = \sup_{X^8}[\beta(x) - I(x)],$$

Then x_N is point of global minimum of Problem 1.

Proof of Theorem 1.8. By subtracting the inequality $\beta \geq \beta_N$ from $\beta - I \leq \beta_N - I_N$ we get $I \geq I_N$ over set X^*. The Theorem 1.8 is proved.

If conditions of Theorem 1.8 is executed only in small sphere around point x_N then x_N is point of local minimum of Problem 1.

The example for illustration of this method (for tests of constrained minimum) will be given in § 4 (remark 4.3). We can get the direction in the set N, if we calcule a gradient of function in N.

The advantegies this method with comparison of gradient method is big steps. When you are in set N, you have not a danger of to get worthier solution than given one. This can substentionaly decrease amount of calculation.

5. Method of β-function for Problems with constrains

A) Assume $I(x)$ is function by its lower estimate over set X. The subset $X^* \neq \emptyset$ is separated from X by functions

$$F_i(x) = 0 \quad i = 1,2,\ldots,k, \quad \Phi_j(x) \leq 0, \quad j = 1,2,\ldots,q, \qquad (1.6)$$

where x - is n-dimentional vector of numerical values.

Take β-function as following (we have a sum for lower index i, j)

$$\beta(x, y) = \lambda_i(x, y)F_i(x) + \omega_j(x, y)\Phi_j(x),$$

where $\lambda_i(x,y)$, $\Phi_j(x, y)$ are functions of x,y, $y \in Y$, $\omega_j(x, y) \geq 0$.

Write J-function

$$J(x, y) = I(x) + \lambda_i(x, y)F_i(x) + \omega_j(x, y)\Phi_j(x). \qquad (1.8)$$

Theorem 1.9. Assume exist $x^* \in X^*$, y is fixed. In other $\bar{x}$ to be a point of global minimum of function $I(x)$ over X^* necessary and enough to exist of function $\beta(x,y)$ such as

1) $J(\bar{x}, y) = \inf_{x \in X} J(x,y)$, 2) $\bar{x} \in X^*$, 3) $\omega_j(x,y) \geq 0$ over X, 4) $\beta(\bar{x},y)=0$, (1.9)

Proof of Theorem 1.9.
Sufficiency. From "1)" of (1.9) we have $I + \lambda_i F_i + \omega_j \Phi_j \geq \bar{I} + \bar{\lambda}_i \bar{F}_i + \bar{\omega}_j \bar{\Phi}_j$.
From this and "4)" (1.9) we get $I + \lambda_i F_i + \omega_j \Phi_j \geq \bar{I}$. Look it inequality over X*. On X* we have
$\lambda_i F_i = 0$, $\omega_j \Phi_j \leq 0$ hence $I(x) \geq I(\bar{x})$. Because $\bar{x} \in X^*$ hence $\bar{x}$ is the point of global minimum of I(x) on X*.
Necessity. (Method of designing). Assume that $x^* \in X$ exists. Design $\beta(x,y)$ following way. Take $\lambda_i \equiv 0$ on X^* and take functions λ_i, $\omega_j \geq 0$ such us $J(x) > m$ on set $X \setminus X^*$. Then we have as the result of our design $J(x^*) = \inf_{x \in X^*} J(x)$, $x^* \in X^*$, $\omega_j \geq 0$, $\bar{\beta} = 0$.
The theorem 1.9 is proved.

Theorem 1.10. (The lower estimation)
Assume y is fixed, $\bar{x}$ is point of minimum (1.8) for conditions $\omega_j(x,y) \geq 0$.
Then $J(\bar{x}, y)$ is lower estimation of function I(x) on X*.

Proof: On set X* we have $\lambda_i F_i \equiv 0$, $\omega_j \Phi_j \leq 0$ (that is $\beta(\bar{x}, y) \leq 0$). Since over X* we have $J(\bar{x}, y) \leq I(x)$.
Theorem is proved.

Likely a common case for β- function we can get the sets
$$M = \{x : \beta \geq \bar{\beta}\}, \quad N = \{x : J + I \leq \bar{J} + \bar{I}\}, \quad P = \{x : \beta \leq \bar{\beta}\}$$
and in this case.

Freedom in choice of y we can use for improvement of estimation and decrease sizes of sets M, N. Remark only that $\bar{x} = \bar{x}(y)$ and for every y corresponding $\bar{x}$ you must find inf J(x,y), x∈X.

Remark:

We can take β-function (1.7) in form
$$\beta(x) = \frac{1}{2} a \sum_{i=1}^{k} F_i^2(x) + \sum_{j=1}^{q} a^{\Phi_j(x)}.$$
It is possible to show for some conditions: [I(x), $\Phi_j(x)$, $F_i(x)$ are continuous, x is compact set, x* is close set and don't contain separated points; x* ∈X* and exist], when $a \to \infty$, we have $\bar{J} \to m$, $\bar{x} = x^*$.

B) Assume $F_i(x) = 0$ in (1.6) absent, i.e. the Problem is
$$I(x) = \min, \quad \Phi_j(x) \leq 0, \quad j = 1,2,...,q \qquad (1.11)$$
For solution of this problem we can use following **algorithm**:

1. Take any functions $\omega(x,y)$ (it's may be less zero) and find the point $\bar{x}(y)$ of global minimum (one may be implicit form $\xi(\bar{x}, y) = 0$) of general numerical function
$$J = I(x) + \sum_{j=1}^{q} \omega_j(x,y) \Phi_j(x) \quad \text{on} \quad X. \qquad (1.12)$$

2. Solve equations
$$\xi(\bar{x}, y) = 0, \quad \omega_j(\bar{x}, y) \Phi_j(\bar{x}) = 0, \quad j = 1,2,...,q \qquad (1.13)$$

3. Select from these solutions such which satisfy inequalities
$$\omega_j(\bar{x},\bar{y}) \geq 0, \quad j=1,2,...,q. \tag{1.14}$$

These are points of global minimum of Problem (1.11) because all request the theorem 1.4 is satisfy.

We can solve (1.13) by different ways. For example, find $\bar{x}$ from equation $\xi(\bar{x},y) = 0$ and substitute in the last equations (1.13)
$$\omega_j(\bar{x}(y),y)\Phi_j(\bar{x}(y)) = 0, \quad j=1,2,...,q \tag{1.15}$$
Find y from this system of equations. Select from these solutions such which satisfy inequalities
$$\omega_j(\bar{x}(y),y) \geq 0, \quad j=1,2,...,q, \tag{1.16}$$
or we can find y from $\xi(\bar{x},y) = 0$ and substitute in the last equations (1.13) and find $\bar{x}$.

6. Application the method of β- functions to linear programming.

The Problem of Linear Programming is
$$I = \sum_{i=1}^{n} c_i x_i = \min, \quad \sum_{j=1}^{n} a_{kj} x_j - b_k \leq 0, \quad k=1,2,...,m \tag{1.17}$$
Here c_i, a_{kj}, b_k are constant.

Take $\omega_j = y_i$. Then equation (1.13) are
$$y_k (\sum_{j=1}^{n} a_{kj} x_j - b_k) = 0, \quad k=1,2,...,m \tag{1.18}$$
$$c_i + \sum_{j=1}^{m} a_{ij} y_j = 0, \quad i=1,2,...,n \tag{1.19}$$

Selective from (1.18) l equations ($l \leq n$, $l \leq m$, $l =$ max) and l variables x_j such that determinant $|a_{kj}| \neq 0$. Find $\tilde{x}_j$ from these l linear equations (1.18) (corresponded $y_k \neq 0$).

If this solution don't satisfy inequalities (1.17), we take l other equations and repeat this procedure (process) while we find $\tilde{x}_j$ which saticfy (1.17). If these equations absent, we take l -1 equations (1.18) and repeat process, than l - 2 equations and so on, while we get l = 0.

If solution, which satisfy (1.17), absent that inequality (1.17) is conflicting (incompatible) and cannot be solved.

Assume that by using this procedure we find the solution $\tilde{x}_j$, that satisfy (1.17). Take in (1.19) all y_j, which don't belong the taken questions (1.18), equal zero and find y from equation (1.19). If all $\tilde{y}_j \geq 0$ then $\tilde{x}_j$ is point of minimum of problem (1.17). If part of $\tilde{y}_j < 0$, then we change corresponded equations (1.18) by other and repeat this process while get all $\tilde{y}_j \geq 0$.

We can suppose that this process makes all $\tilde{y}_j \geq 0$. Inequality $\tilde{y}_j \geq 0$ means that anti-gradient has direction into internal of the corresponding constraints. Because our problem and constrains are linear, anti-gradient, which has direction into constrains, will has this direction in any point of corresponding hyper plate (1.17). It means that this procedure will increase the amount of $y_j \geq 0$.

Example 1.4.

Find minimum of Problem
$$I = x_1 + x_2, \quad -x_1 \leq 0, \quad -x_2 \leq 0, \quad x_1 - 1 \leq 0, \quad x_2 - 1 \leq 0. \tag{1.20}$$

The equations (1.18),(1.19) are
$$-y_1 x_1 = 0, \quad y_3(x_1 - 1) = 0, \quad 1 - y_1 + y_3 - 0,$$
$$-y_2 x_2 = 0, \quad y_4(x_2 - 1) = 0, \quad 1 - y_2 + y_4 = 0. \quad (1.21)$$

Chose equations $x_1 - 1 = 0$, $x_2 - 1 = 0$. From solution of them we have $\tilde{x}_1 = 1$, $\tilde{x}_2 = 1$. They satisfy (1.20). From the first column of (1.21) we get $y_1 - y_2 = 0$, and from the last column (1.21) we find $y_3 = y_4 = -1$. Inequality $y_i \geq 0$ is not satisfied. Change equalities by others $\tilde{x}_1 = 0$, $\tilde{x}_2 = 0$. We get $\tilde{y}_1 = \tilde{y}_2 = 1 > 0$. Hence $\tilde{x}_1 = \tilde{x}_2 = 0$ is point of the global minimum.

Example 1.5.
Find point of global minimum in Problem $I = -x_1 - x_2$, $x_1 + x_2 \leq 0$.
Solution. Write equations (1.18),(1.19) $y(x_1 + x_2) = 0$, $-1 + y$,
From $x_1 + x_2 = 0$ we get $\tilde{x}_1 = -\tilde{x}_2$. From $-1+y = 0$ we get $y = 1 > 0$. Sence any $\tilde{x}_1 = -\tilde{x}_2$ is optimal.

7. Application of method β-function to quadratic programming.

This problem is following:

$$I = \sum_{j=1}^{n}\sum_{i=1}^{n} c_{ij} x_i x_j, \quad \sum a_{kj} x_j - b_k \leq 0, \quad k = 1,2,...,m. \quad (1.22)$$

Assume that quadratic form in function (1.22) is positive. If don't consider constraints in (1.22), it is obvious the point of minimum in this problem is $x_j^* = 0$. If this point satisfy inequalities in (1.22), the process of solution is finished. In particular, we have this case when all $b_k \geq 0$. We consider not triviality case. Take $\omega_j = y_j$.
Equations (1.13) and (1.14) are:

$$y_k(\sum_{j=1}^{n} a_{kj} x_j - b_k) = 0 \quad i,k = 1,2,...,n; \quad \sum_{j=1}^{n} c_{ij} x_j + \sum_{j=1}^{m} y_l a_{jk} = 0, \quad y_k = 0. \quad (1.23)$$

Later procedure is analogous of the Linear Programming.

Example 1.6.

Problem are:
$$I = 0.5 x_1^2 + 0.5 x_2^2, \quad -x_1 - x_2 + 1 \leq 0, \quad x_1 - 1 \leq 0, \quad x_2 - 1 \leq 0. \quad (1.24)$$

The equations (1.23)
$$y_1(-x_2 - x_1 + 1) = 0, \quad y_2(x_1 - 1) = 0, \quad y_2(x_2 - 1) = 0$$
$$x_1 - y_1 + y_2 = 0, \quad x_2 - y_1 + y_2 = 0 \quad (1.25)$$

Take the 2-nd and 3-rd equations. We get $\tilde{x}_1 = \tilde{x}_2 = 1$. The inequalities (1.24) are satisfied, but from two the last equations (1.25) for $y_1 = 0$ we have $\tilde{y}_2 = \tilde{y}_3 = -1$. It is contrary the request $\tilde{y}_i \geq 0$.

Take the 1-st equation in (1.25). We have $\tilde{x}_2 = 1 - \tilde{x}_1$. Solve it together with equations $\tilde{x}_1 - \tilde{y}_1 = 0$, $\tilde{x}_2 - \tilde{y}_1 = 0$ we get $\tilde{x}_1 = \tilde{x}_2 = 1/2$, $\tilde{y}_1 = \tilde{y}_2 = 1/2 > 0$. Hence $x_1 = x_2 = 1/2$ is point of global minimum.

§2. Method of combining of the extremes.

Let us to have the problems:

Problem 1 $\quad I(x^*) = \inf I(x), \quad x \in X^*;$

Problem 2 $\quad J(\bar{x}) = \inf[I(x) + \beta(x)], \quad x \in X$;

Problem 3 $\quad \beta(\hat{x}) = \sup \beta(x), \quad x \in X.$

Assume that all points $x^*, \bar{x}, \hat{x}$ are exist.

Theorem 2.1. *Let $X = X^*$, then for every couple $(\bar{x}_i, \hat{x}_i)$ which satisfy the condition $\bar{x}_i = \hat{x}_i$ we have*
$$\bar{x}_i = \hat{x}_i = x_i^*.$$

Proof. Let $\bar{x}_i = \hat{x}_i$ Then
$$\inf J(x) - \sup \beta(x) = J(\bar{x}_i) - \beta(\bar{x}_i) = I(\bar{x}_i) - \beta(\bar{x}_i) - \beta(\bar{x}_i) = I(\bar{x}_i).$$

But with other side from Theorem 1.2 we have $\inf J(x) - \sup \beta(x) \leq \inf I$. That is $I(\bar{x}_i) \leq I(x^*)$. As x^* is point of global minimum and $X = X^*$ hence must be only $I(\bar{x}_i) = I(x_i^*)$. As far as $\bar{x}_i$ and x_i^* exist we can find the point of minimum x_i^* such that $\bar{x}_i = x_i^*$. Theorem 2.1 is proved.

Theorem 2.2. *Let $X = X^*$. If exist at least one of the couple $(\bar{x}_i, \hat{x}_i)$ such that $\bar{x}_i = \hat{x}_i$, then in every point x_i^* we have*
1) $x_i^* = \hat{x}_i$, 2) $x_i^* = \bar{x}_i$.

Proof. 1. Assume the contrast: $\bar{x}_i \neq x_i^*$. Than summarize $I(\bar{x}_i) = I(x_i^*)$ and $\beta(x_i^*) < \beta(\bar{x}_i) = \beta(\hat{x}_i)$ we get $J(x_i^*) < J(\bar{x}_i)$. This contrasts $J(\bar{x}_i) = \inf J(x)$.

2. Add $J(\bar{x}_i) = J(x_i^*)$ and $\beta(x_i^*) = \beta(\bar{x}_i) = \beta(\hat{x}_i)$ we get $J(x_i^*) = J(\bar{x}_i)$, hence $x_i^* = \bar{x}_i$. Theorem 2.2 is proved.

From Thorems 2.1, 2.2 we have

Consequence:

If we want to find all points of minimum of Problem 1 it necessary and sufficiently to find all corresponding couple $(\bar{x}_i, \hat{x}_i)$.

We shall call the Problems 1 and 2 <u>equivalents</u> if all correspondent points of minimum of these Problems are coincided.

From Theorem 2.2 we have:

1. For equivalence of Problems 1, 2 is sufficient to exist one couple such that $\bar{x}_i = \hat{x}_i$.
2. Let exist β-functional and although one of couple $(\bar{x}_i, \hat{x}_i)$ such that $\bar{x}_i = \hat{x}_i$.

Then any points of minimum of Problem 2 and point of maximum of Problem 3 is point of minimum of Problem 1, and back, any point of minimum of Problem 1 is point of minimum of Problem 2 and point of minimum of Problem 3.

Remarks:

1. If $\beta(\bar{x}) = 0$, then $\inf J(x) = \inf I(x)$.
2. If $\bar{x} = \hat{x}$, then the lower estimate (1.1) in §1 coincide with infinum of the functional $I(x)$.

From consequence 1 §2 we have the following

Algorithm 3. (Method of combining the extremes)

Let us take some bounded functional $\beta(x,y)$ where y is an element of the set Y. We solve this problem
$$\inf[I(x)+\beta(x,y)], \quad x \in X^*$$
and find the point of minimum
$$\bar{x}_1 = \bar{x}_1(y)$$
From
$$\sup \beta(x,y)$$
we find
$$\bar{x}_2 = \bar{x}_2(y)$$
After this we equate
$$\bar{x}_1(y) = \bar{x}_2(y) \quad (2.1)$$
and from this equation of the combination of extreme we find the roots y_i.

These roots are the points of minimum for Problem 1:
$$\bar{x} = \bar{x}_1(y_i) = \bar{x}_2(y_i)$$

Since the Problem of finding of global minimum is reduced to Problem of finding at least one root of equation of the combination of extremes (2.1).

The exist and difficulty of finding of roots dipend from chouse of β-functional, from freedom of its deformation, which give the "y" relation.

Note that is differ from the regular method of finding of minimum. In the usual method we take partial derivatives, equal its zero, get the set equation and from them we find only the stationary (extreme) points. They may be points local minimum, maximum, or inflection. By this method we find points of global minimum.

Thus we find the connect two various (different) problems.

The existence of solution in equation of the combination of extremes is sufficient condition for the existence of absolute minimum of functional in Problem 1.

The mathematic has good achievements in the field of existence of solution of equations. And equation (2.1) give connection between these problems and give some opportunity in solving of optimals problems.

Note also that equation (2.1) not requests that functional was continuous and differential function, hence it has wider domain for application.

If point of minimum cannot be get in explicit form than we can write this equation in form
$$\varphi_1(x,y) = 0, \quad \varphi_2(x,y) = 0, \quad (2.1')$$
where function φ_1, φ_2 are got from
$$\inf_X J(x,y), \quad \sup_X \beta(x,y).$$

Example 2.1. Find a point of minimum of functional
$$I = 2x^4 + x^2 - 2x + 1, \quad -\infty < x < \infty$$

<u>Solution</u>: Use algorithm 3. Take

$$\beta = -yx^2 + 2x.$$

Than

$$J = I + \beta = 2x^4 + (1-y)x^2 + 1.$$

Denote $x^2 = w$ and substitute in J:

$$J = 2w^2 + (1-y)w + 1.$$

Find point of minimum this functional $J'_w = 4w + (1-y) = 0$, $\overline{w} = \overline{x}_1^2 = \frac{1}{4}(y-1)$ and point of maximum functional β. $\beta(x) = -yx^2 + 2x$, $\beta'_x = -2yx + 2 = 0$, $\overline{x}_2 = 1/y$.

Equate $\overline{x}_1$ to $\overline{x}_2$

$$\overline{x}_1^2 = \overline{x}_2^2 \quad \frac{1}{4}(y-1) = \frac{1}{y^2}, \quad y^3 - y^2 - 4 = (y-2)(y^2 + y + 1)$$

This equation has only alone root $\overline{y} = 2$. Since $\overline{x} = \frac{1}{y} = \frac{1}{2}$.

§3. Remark about γ-functional

A) Let us take

$$\beta(x) = [\gamma(x) - 1]I(x) \tag{3.1}$$

then

$$J(x) = I(x)\gamma(x).$$

This form of common functional is sometimes more comfortable because we can chouse the multiplier to I(x) which make J(x) simpler.

Using our results about β-functional for this case we get following:
If X = X* and we finding the point of global minimum Problem 2:

$$\inf_X J(x) = \inf_X [I(x)\gamma(x)] \tag{3.2}$$

than

1) Set

$$M = \{x : J - I \geq \overline{J} - \overline{I}, \quad x \in X\}$$

contains the point of global minimum of Problem 1;

2) Set

$$N = \{x : I\gamma + I \leq \overline{I}\overline{\gamma} + \overline{I}, \quad x \in X\}$$

contains the better or same solutions than $\overline{x}$ (that is over N, we have $I(x) \leq I(\overline{x})$);

3) Set

$$P = \{x : J - I \le \bar{J} - \bar{I}, \quad x \in X\}$$

contains the worse or same solutions than $\bar{x}$ (that is over P, we have $I(x) \ge I(\bar{x})$).

All these statement follow from (3.1) and Theorem 1.1.

Lower estimate (from Theorem 1.3 and (3.1)) look as

$$I(x) \ge \inf_X J - \sup_X (J - I). \qquad (3.3)$$

Condition of equivalence of Problem 1 and 2 (theorem 2.1) in this case ($X=X^*$) is:

$\bar{x}$ and $\hat{x}$, which are founded from problems

$$\inf_X J(x) \quad \text{and} \quad \sup_{X^g}[J(x) = I(x)],$$

must equal respectively.

Algorithm 3 (Method of combining the extremes) is used for this case without change.

B) However for this case we get some new results.

Let define functional $\gamma(x,y) \ne 0$ over set $X \times Y$. We call it as γ-functional. Take functional

$$J(x, y) = I(x)\gamma(x, y)$$

Theorem 3.1.

Assume $X = X^*$, $\bar{x}$ is point of global minimum of Problem 2:

$$\inf J(x), \quad x \in X, \quad \text{where} \quad J = I(x)\gamma(x),$$

Then:

1) Set

$$P = \{x : 0 < \gamma \le \bar{\gamma}\}$$

contains worth or same solutions of Problem 1 (that is $I(x) \ge I(\bar{x})$ over P);

2) Set

$$N = \{x : 0 > \bar{\gamma} \ge \gamma\}$$

contains better or same solution of Problem 1 (that is $I(x) \le I(\bar{x})$ over N);

3) The point of global minimum is in set $M = X \setminus \beta$, where $\overset{o}{P} = \{x : 0 < \gamma < \bar{\gamma}\}$.

Proof: 1. From inequalities $I\gamma \ge \bar{I}\bar{\gamma}$, $0 < \gamma \le \bar{\gamma}$ we have $I \ge \bar{I}\bar{\gamma}/\gamma$, $\bar{\gamma}/\gamma \ge 1$. That is $I \ge \bar{I}$.

2. From inequalities $I\gamma \ge \bar{I}\bar{\gamma}$, $0 > \gamma \ge \bar{\gamma}$ we get $I \le \bar{I}\bar{\gamma}/\gamma$, $\bar{\gamma}/\gamma \le 1$. That is $I \le \bar{I}$.

3. Because $X = M+P$ and $M \cap \overset{o}{P} \ne 0$, we have $M = X - \overset{o}{P}$. Theorem is proved.

Theorem 3.2. Assume $\sup\limits_{X} \gamma > 0$. Then we have the lower estimation

$$I(x) \geq \frac{\bar{J}}{\sup \beta} \quad \text{on} \quad X. \tag{3.4}$$

If $\sup\limits_{X} \gamma(x, y) > 0$ for $\forall y \in Y$, we have the lower estimate

$$I(x) \geq \sup\limits_{Y} \left(\frac{\bar{J}}{\sup\limits_{X} X} \right) \tag{3.4}'$$

<u>Proof:</u> 1) For written conditions from $I\gamma \geq \bar{I}\bar{\gamma}$ we got $I \geq \bar{J}/\gamma$ and $I \geq \bar{J}/\sup\limits_{X} \gamma$.

2) Take maximum of this estimate by y, we get expression (3.4)'.

<u>Example 3.1.</u> Find the lower estimate for functional

$$I = (x^2 - \cos x + 1)e^{(x-1)^2} \quad -\infty < x < \infty.$$

Take $\gamma = e^{-(x-1)^2}$. Then $J = x^2 - \cos x + 1$.

It is obvious the point of minimum this functional $\bar{x} = 0$, $\bar{\gamma} = 1 > 0$, $\sup\limits_{X} \gamma = 1$.

Use the estimate (3.4) we get $I(x) \geq 0$. But for $x = 0$ we have $I(0) = 0$. That way $x = 0$ is point of global minimum.

§4. Application β- function to the multi-variables nonlinear problems of constrained optimization and to problems described by regular differential equations.

A) The first problem is following. Find minimum of functional

$$I = f_0(x), \tag{4.1}$$

Where x-n-dimensional vector, which satisfy independent equations

$$f_i(x) = 0, \quad i = 1, 2, \ldots, m \leq n. \tag{4.2}$$

Functions f(x) is defended in the open domain n-dimensional vector of space X. The admissible set X* separate from X by equations (4.2).

Let us take some functional $\beta(x)$, such that to find

$$\inf[f_0(x) + \beta(x)] \quad \text{on} \quad X^*.$$

It is easier to solve.

Then from solution of Problem 2 in accordance with thorems of §1 we get the following information about Problem 1:

1) The point of global minimum is in set $M = \{x : \beta(x) \geq \beta(\bar{x})\}$;

2) The set $N = \{x: 2f_0 + \beta \leq 2\bar{f}_0 + \bar{\beta}\}$ contains better ans same solutions (that is $f_0(x) \leq f_0(\bar{x})$ on N);
3) The set $P = \{x: \beta(x) \leq \beta(\bar{x})\}$ contains worth and same solutions (that is $f_0(x) \geq f_0(\bar{x})$ on P;
4) If $X = X^* \subseteq P$, that $\bar{x}$ is point of global minimum of problem 1 (consequence 3 of §1).

Let us assume we widen the set X^* for simplification of solution. For example, we rejecte the part of constrains (4.2). Then we have

5) If $X^* \cap M = \emptyset$, than $J(\bar{x})$ is lower estimation $f_0(x)$ on X^* (consequence 5, §1).

It is more comfortable some times to take the suitable $J(x)$ at first and find the point minimum of problem inf $J(x)$ on X^*.

Then the corresponding sets will be (from theorem 1.1')
$$M = \{x: J - I \geq \bar{J} - \bar{I}\}, \quad N = \{x: J + I \leq \bar{J} + \bar{I}\}, \quad P = \{x: J - I \leq \bar{J} - \bar{I}\}.$$

If we solve the problem $\beta(\hat{x}) = \sup \beta(x)$ on $X \supseteq X^*$ we get the additional lower estimate
$$f_0(x) \geq f_0(\bar{x}) + \beta(\bar{x}) - \beta(\hat{x}),$$

(theorem 1.3) and set
$$M = \{x: f_0 + \beta \leq \hat{f}_0 + \hat{\beta}\}, \quad N = \{x: \beta - f_0 \geq \hat{\beta} - \hat{f}_0\}, \quad P = \{x: f_0 + \beta \geq \hat{f}_0 + \hat{\beta}\}.$$

(theorem 1.4).

Take series β we can get the solution of one from Problems of §1 or to facilitate the solution of Problem 1.

The example for case $X^* = X$ was over (see Examples 1.1-1.3). Explain by simple examples (how you can apply the method β-functional for case, when $X^* \neq X$ that is problem with constrains.

<u>Example 4.1</u>. Find minimum of functional
$$I = x \quad \text{on} \quad x^2 + y^2 - 1 = 0.$$

Take any admissible point, for example $\bar{x}_0 = 1$, $\bar{y}_0 = 0$ and $J(x)$ functional as
$$J_1 = (x - x_0)^2.$$

The point of minimum of this functional is obvious $\bar{x} = x_0$. The set M, containing the point of global minimum, is
$$J_1 - I \geq \bar{J}_1 - \bar{I}, \quad \text{that is} \quad (x-1)^2 - x \geq -1 \quad \text{or} \quad |x - 3/2| \geq 3/2$$

The boundaries of this inequality together with admissible subset (circle) draw on fig.1.3a. We see the point of absolute minimum is in left half of circle.

Take now the admissible point $\bar{x}_0 = -1$, $y = 0$ and J-functional in more common case as
$$J_2 = c(x - x_0)^2, \quad c > 0.$$

Then M set is
$$cx^2 + 2cx + c - x \geq 1.$$

Take c = 0.5. Then we get $|x| \geq 1$ (fig. 1.3b).

Set *M* contain only two admissible point : $x_1=1$ and $x_2=-1$. But point $x_1=1$ from the J_1 cannot be the point of absolute minimum. Since the point of global minimum is $\bar{x}=-1$, $\bar{y}=0$.

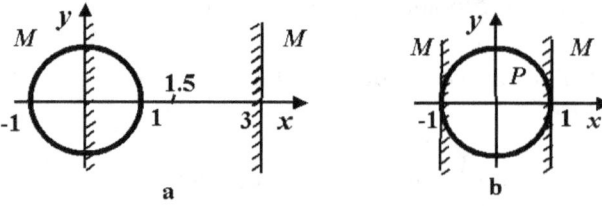

Fig. 1.3

<u>Example 4.2.</u> Find the point of global minimum of functional with constrain

$$I = x^2 - x + y^2 - 2y + 1, \quad y - \ln\left(x \pm \sqrt{x-1}\right) = 0.$$

Take *J* functional

$$J = (x - x_0)^2 + (y - y_0)^2.$$

The set *M* is separated by inequality $J - I \geq \bar{J} - \bar{I}$, or $2y(1 - y_0) \geq (2x_0 - 1)x + a$, where

$$a - x_0 - 2x_0^2 + 2y_0 - 2y_0^2.$$

Take the admissible point $x_0 = -1$, $y_0 = 0$. Then

$$M = \left\{x, y : y \geq \frac{1}{2}x - \frac{1}{2}\right\} \quad \text{(Fig.1.4)}.$$

From drawing we see *M* is small domain and find the point of global minimum no difficult.

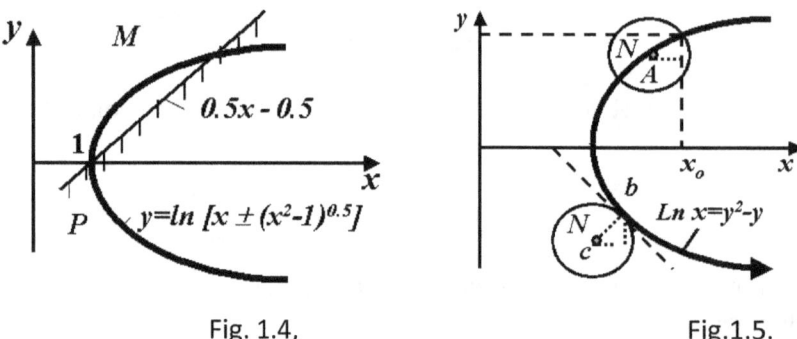

Fig. 1.4,   Fig.1.5.

<u>Example 4.3.</u> Given functional and constrains is

$$I = 2x + 2y, \quad \ln x = y^2 - y$$

Take

$$J = (x - x_0)^2 - (y - y_0)^2,$$

where couple x_0, y_0 is admissible point.

The set N is separated with according Theorem 1.1 by inequality $J + I \leq \bar{J} + \bar{I}$, that is

$$[x - (x_0 - 1)]^2 + [y + (y_0 - 1)]^2 \leq 2.$$

This is interior of the circle (fig.1.5).

Assume that a center of this circle is located in the point A. The set N intersect with admissible curve In $x = y^2 - y$. If we take a point x_0, y_0 from this intersection, we will descent along this curve whole the set N become by point. This take place in point B, where the tangent to admissible curve has the angle -45⁰ (because the center of the circle is located from point x_0, y_0 from -1, -1, that is the angle +45⁰, (fig. 1.5). Any moving from this point will return us to it.

May be shown that the point B is the point of global minimum.

Take into consideration when we have used the methods of β-functional (Chapter 1) we have not used in continuously and differ of functional (4.1) and constancies (4.2) unlike from known methods (for example, theory of extreme functions).

B) Consider how we can apply the methods given in §1 to optimization problems are described by regular differential equations. Below we write the statement of problem, which we widely use in future.

Assume that the moving of object is described by set of independent differential equations

$$\dot{x}_i = f_i(t, x, u), \quad i = 1, 2, ..., n, \quad t \in T = [t_1, t_2], \tag{4.3}$$

where $x(t)$ is n-dimensional continually piece-differential vector-function of the phase coordinates, $x \in G(t)$; $u(t)$ is n-dimensional function which continuous on T except the limited number of point where it can have discontinuities of the 1-st form, $u \in U$ is an indepeded variable. Boundary values t_1, t_2 is given, $x(t_1) \in G(t_1)$, $x(t_2) \in G(t_2)$.

The aim function is

$$I = F(x_1, x_2) + \int_{t_1}^{t_2} f_0(t, x, u) dt, \quad x_1 = x(t_1), \quad x_2 = x(t_2). \tag{4.4}$$

Functions $F(x_1, x_2)$, $f_i(t, x, u)$, $i = 0, 1, ..., n$ are continuous over $T \times G \times U$. Set of continuous, almost everywhere differentiable functions $x(t) \in G(t)$ we denote D. Set of pies-continuous functions $x(t) \in U$, we denote V. Set of couple $x(t)$, $u(t)$ which satisfy these requirements and almost everywhere comply with equations (4.3) we shall call **admissible** and denote Q, $Q \subset D \times V$.

Consider the problems:

a) Find the coiple $u^*(t)$, $x^*(t) \in D$, which give the minimum of function (4.4) (Traditional statement).
b) Find sup-set $N \subset G \times U \times T$ such that any admissible curve from N we have $I(x) \leq c$, where c is constant.
c) Find the lower estimate of $I(x)$ over Q.

Take the function $\int_{t_1}^{t_2} \beta(t, x, u) dt$, where $\beta(t, x, u)$ is a definite and continuous function on $T \times G \times U$.

Theorem 4.1. Let us assume that $F \equiv 0$ and Problem 2 is solved. That mean
$$J(\bar{x}, \bar{u}) = \inf J(x, u) \quad \text{on} \quad Q,$$
where $J = \int_{t_1}^{t_2} [f_0(t, x, u) + \beta(t, x, u)] dt$

Then:

1) Set
$$N = \{t, x, u : 2f_0 + \beta \leq 2\bar{f}_0 + \bar{\beta}, \quad t \in T\}$$
contains the same or better solutions of Problem 1.

2) Set
$$P = \{t, x, u : \beta \leq \bar{\beta}, \quad t \in T\}$$
contains the same or worse solutions of Problem 1.

Proof: 1. On set Q from N we have
$$\int_{t_1}^{t_2}(2f_0 + \beta)dt \leq \int_{t_1}^{t_2}(2\bar{f}_0 + \bar{\beta})dt.$$

Subtract from this inequality following
$$\int_{t_1}^{t_2}(f_0 + \beta)dt \geq \int_{t_1}^{t_2}(\bar{f}_0 + \bar{\beta})dt \qquad (4.5)$$

we get over Q from N
$$\int_{t_1}^{t_2} f_0 dt \leq \int_{t_1}^{t_2} \bar{f}_0 dt.$$

2. By analogy with above, subtract from inequality
$$\int_{t_1}^{t_2} \beta dt \leq \int_{t_1}^{t_2} \bar{\beta} dt$$

the inequality (4.5) we get over Q from P
$$\int_{t_1}^{t_2} f_0 dt \geq \int_{t_1}^{t_2} \bar{f}_0 dt$$

The Theorem 4.1 is proved.

Sets N, P not empty. They contain at least one curve from Q. This curve is $\bar{x}(t), \bar{u}(t) \in Q$.

If we solve the additional problem
$$\sup_Q \int_{t_1}^{t_2} \beta dt,$$

we get additional information about sets N, P and lower estimate. It is following

Theorem 4.2. *Let us assume* $F \equiv 0$ *and solved the Problem*
$$\sup \int_{t_1}^{t_2} \beta(t, x, u) dt \quad \text{on} \quad Q$$

Then
1) Set
$$N = \{t, x, u : \beta - f_0 \geq \hat{\beta} - \hat{f}_0, \quad t \in T\}$$
contains the same or better solutions:

2) Set
$$P = \{t, x, u : f_0 + \beta \leq \hat{f} + \hat{\beta}, \quad t \in T\}$$
contains the same or worse solutions.

Here $\hat{f}_0 = f_0(t, \hat{x}, \hat{u})$, $\hat{x}(t)$, $\hat{u}(t)$ is curve of extreme $\sup \int_{t_1}^{t_2} \beta(t)$ on Q.

Proof: 1. Over Q from N we have

$$\int_{t_1}^{t_2}(\beta - f_0)dt \geq \int_{t_1}^{t_2}(\hat{\beta} - \hat{f}_0)dt$$

Subtract from this inequality the following

$$\int_{t_1}^{t_2}\beta dt \leq \int_{t_1}^{t_2}\hat{\beta}dt,$$

we get

$$\int_{t_1}^{t_2}f_0 dt \leq \int_{t_1}^{t_2}\hat{f}_0 dt.$$

2. By analogy, subtract $\int_{t_1}^{t_2}\beta dt \leq \int_{t_1}^{t_2}\hat{\beta}dt$ from

$$\int_{t_1}^{t_2}(f_0 + \beta)dt \geq \int_{t_1}^{t_2}(\hat{f}_0 + \hat{\beta})dt$$

we get

$$\int_{t_1}^{t_2}f_0 dt \geq \int_{t_1}^{t_2}\hat{f}_0 dt.$$

The Theorem 4.2 is proved.

Theorem 4.3. *(Lower estimation).*
Assume $F \equiv 0$, the ends of $x(t)$ are fixed, $\beta(t,x,u)$ is defined and bounded on $G \times U \times T$. Then there is lower estimate of Problem 1:

$$I(x,u) \geq \int_T [f_0(t,\bar{x},\bar{u}) + \beta(t,\bar{x},\bar{u}) - \beta(t,\hat{x},\hat{u})]dt \qquad (4.6)$$

Proof: Subtract $\int_T \beta dt \leq \int_T \sup \beta dt$ from inequality

$$\int_T (f_0 + \beta)dt \geq \int_T (\bar{f}_0 + \bar{\beta})dt$$

we get (4.6). The theorem 4.3 is proved.

Consequence 1: Couple $\bar{x}, \bar{u}$ is curve of absolute minimum of Problem 1 over set N.

Consequence 2: If set $P \supseteq T \times G \times U$ (or accessible) then $\bar{x}, \bar{u}$ (or $\hat{x}, \hat{u}$) is curve of global minimum of problem 1 over Q.

Similar results we can get for case, when $F \neq 0$ and ends of $x(t)$ can move.

Example 4.4. Assume the problem is described by conditions:
$$I = \int_0^1 (x^2 + e^u)dt, \quad \dot{x} = u, \quad |u| \leq 1, \quad x(0) = 1, \quad x(1) = 0.$$

Use the theorem 4.1. Take $\beta = -e^{+u}$. We get the problem
$$I = \int_0^1 x^2 dt, \quad \dot{x} = u, \quad |u| \leq 1, \quad x(0) = 1, \quad x(1) = 0.$$

Its solution is $\bar{x} = -t, \quad \bar{u} = -1, \quad 0 \leq t \leq 1$.

Find set P: $\beta \le \bar{\beta}$. That is $e^u \ge e^{-1}$, $u \ge 1$.

But value u < -1 is not acceptable. Since P is cover all admissible set points t,x,u. That way $\bar{x} = -t$.

Is the curve of global minimum (see Consequence 2).

Example 4.5. Find of minimum in problem

$$I = \int_0^2 (|x| + 0.5x^2) dt, \quad \dot{x} = u, \quad |u| \le 1, \quad x(0) = 1, \quad x(2) = 0.$$

We have here undifferentiated function in integral. Known methods us variational calculation or principle of maximum are not been used.

Change this problem following "good" (easy) problem:

$$L = \int_0^2 0.5x^2 dt, \quad \dot{x} = u, \quad |u| \le 1, \quad x(0) = 1, \quad x(2) = 0$$

and find $\sup_{x(t)} L$.

The solution is shown in Fig. 1.6.

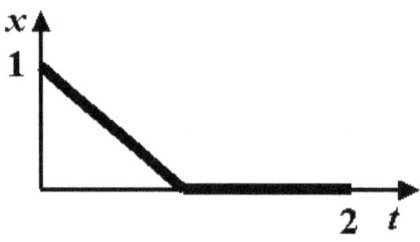

Fig. 1.6. For Example 4.5.

By according the theorem 4.2

$$P = \{x : |x| \ge |\bar{x}|\},$$

that means set P cover all accessible domain. Since abtained, solution is curve of global minimum of Problem 1.

5. Method of β- function in minimizing sequences

A) The sequence $\{x_s\}$ such that $I(x_s) \xrightarrow[s \to \infty]{} \inf I(x)$ on the set X^* is named as a minimizing sequence (for Problem 1).

We must design these sequence in a successive approximation methods and in case, when extreme is absent in an allowable (admissible) subset.

Theorem 5.1. Assume $\beta(x) \le 0$ on X^* and there exist sequence $\{x_s\} \in X^*$ such, that

$$J(x_s) \to \inf J \quad \text{for} \quad s \to \infty \quad \text{on} \quad X. \tag{5.1}$$

Then: 1) $I(x_s) \xrightarrow[s \to \infty]{} m = \inf I(x)$ on X^*;

2) *Any sequence $\{x_s\} \in X$, which satisfy (5.1) or $I(x_s) \to \inf_X J$, minimize I(x) on X*, minimize and J(x) on X.*

<u>Proof:</u> 1. Because $\beta(x) \leq 0$ on X*, we have $\inf_X J \leq I(x)$. That is $\inf_X J \leq \inf_{X^*} J$. From $\{x_s\} \in X^*$ and (5.1) we have that
$$\inf_X J = \inf_{X^*} I. \qquad (5.2)$$

That is $I(x_s) \to m$.

2. From (5.1) and (5.2) we have the statement 2 of the theorem.

3. From $I(x_s) \xrightarrow[s\to\infty]{} m$ and (5.2) we have $J(x_s) \to \inf J$ for $s \to \infty$ on X. Theorem is proved.

Remark. The requirement $\beta(x) \leq 0$ on X* of the theorem 5.1 we can change by the requirement $\sup_{X^*} \beta \leq 0$ on X^* because from $\sup \beta \leq 0$ on X* we have $\beta(x) \leq 0$ on X*.

Theorem 5.2. *Assume there exist the sequence $\{x_s\} \in X^*$ such that*
$$J(x_s) \xrightarrow[s\to\infty]{} \inf J(x) \text{ on } X \text{ (or } X^*\text{)} \quad \text{and} \quad \beta(x_s) \to \sup \beta \text{ on } X \text{ (or } X^*\text{)} \qquad (5.3)$$

Then this sequence is minimized.

<u>Proof</u>: From $I(x_s) + \beta(x_s) \to \inf J$ and $\beta(x_s) \to \sup \beta$ we get that $I(x_s) \to \inf J - \sup \beta$. Because $I(x_s) \geq \inf J - \sup \beta$ and there exist $\{x_s\} \in X^*$ we have $I(x_s) \to m = \inf J - \sup \beta$.

Q.E.D.

Remark: From (1.1) and (1.1') we see that X and X* in (5.3) we can take in any combinations.

B) Let us consider a case now, when we have both a sequence of elements $\{x_s\}$ and a sequence of functions $\{\beta_i(x)\}$.

Theorem 5.3. *In order that a sequence $\{x_s\} \in X^*$ minimize function I(x) on set X*. It is sufficient that there exist a sequence of functions $\{\beta_i(x)\}$ such that*

(1) $\beta_i(x) \leq 0$ over X* for all i;
(2) *There exist numbers* $q_i = \inf_X J_i$, $q = \lim_{i\to\infty} q_i$;
(3) $J(x_s) \to q$ or $I(x_s) \to q$ if $s \to \infty$.

This theorem may be proved easy, because q = inf I over set X*.

From theorems 5.1, 3.3 we have next *statement*:

If there exist one sequence which satisfy theorem 5.3 then any other sequence which belong to set X, $\{x_s\} \in X$ and satisfy the condition $I(x_s) \to q$ or $J(x_s) \to q$ is minimize for Problem 1.

Appendix to Chapter 1.

1. **Operations with signs** inf **and** sup.
   Below there shown the characteristics of signs **inf** and **sup,** which can be useful for solution of problems. The proof is simply and no given. We assume that are shown constrains have place in domain of definition of function.

   1. $\inf[-f(x)] = -\sup f(x), \quad \sup[-f(x)] = -\inf f(x)$.
   2. $\inf cf(x) = c \inf f(x) \quad \text{if} \quad c = const > 0;$
      $\inf cf(x) = -c \inf f(x) \quad \text{if} \quad c = const < 0.$
   3. $\inf[c + f(x)] = c + \inf f(x)$,
   4. $\inf \dfrac{1}{f(x)} = \dfrac{1}{\sup f(x)} \quad \text{if} \quad f(x) \neq 0.$

   5. If $\bar{x}(t)$ can have breaks and $f(t, \bar{x}(t))$ has integrality then

   $$\inf_{x(t)} \int_{t_1}^{t_2} f[t, x(t)] dt = \int_{t_1}^{t_2} \inf_{x} f(t, x) dt .$$

   6. Assume $f(\varphi)$ is monotone function, $\partial f / \partial \varphi$ is continuous. Then

   $$\inf f[\varphi(x)] = f[\inf_{X} \varphi(x)] \quad \text{if} \quad \partial f / \partial \varphi > 0,$$

   $$\inf f[\varphi(x)] = f[\sup_{X} \varphi(x)] \quad \text{if} \quad \partial f / \partial \varphi < 0.$$

   Consequences:

   Functions of single variable
   a) $\inf f^{2n}(x) = [\inf f(x)]^{2n}$, if $f(x) \geq 0, \quad m \geq 0 - \text{intenger}$
      $\inf f^{2n}(x) = [\inf f(x)]^{2n}$, if $f^{2n-1}(x) > 0,$

   $\inf f^{2n}(x) = [\sup f(x)]^{2n}$, if $f^{2n-1}(x) < 0.$
   b) $\inf f^{2n+1}(x) = [\inf f(x)]^{2n+1}$,
   c) $\inf \log_a f(x) = \log_a \inf f(x), \quad \text{if} \quad a > 1.$
      $\inf \log_a f(x) = \log_a \sup f(x), \quad \text{if} \quad 0 < a < 1.$
   d) $\inf a^{f(x)} = a^{\inf f(x)}, \quad \text{if} \quad a > 1.$
      $\inf a^{f(x)} = a^{\sup f(x)}, \quad \text{if} \quad 0 < a < 1.$
   e) $\inf \sin f(x) = \sin \inf f(x) \quad \text{in} \quad \text{domain} \quad (-0.5\pi \leq x \leq 0.5\pi).$
   f) $\inf \cos f(x) = \cos \sup f(x) \quad \text{in} \quad \text{domain} \quad (0 \leq x \leq \pi)$
   g) $\inf \operatorname{arctg} f(x) = \operatorname{arctg} \inf f(x) \quad \text{in} \quad \text{domain} \quad (0 \leq x \leq \pi)$
   g) $\inf a \tan f(x) = a \tan \inf f(x). \quad \text{in} \quad \text{domain} \quad (-0.5\pi \leq x \leq 0.5\pi), a > 0.$

   h) $\inf \tan f(x) = \tan \inf f(x), \quad \text{if} \quad |f(x) < \pi/2|.$

i) $\inf_t \sqrt{f(x)} = \sqrt{\inf f(x)}$.

j) $\inf_t \dfrac{df(x)}{dt} = \dfrac{d}{dt} f(x)\Big|_{\bar{t}}$ in domain $f''(t) > 0$. Here $\bar{t} = \arg\inf_t f(t)$.

k) $\inf_t \dfrac{df(x)}{dt} = \dfrac{d}{dt} f(x)\Big|_{\bar{t}}$ in domain $f''(t) < 0$. Here $\bar{t} = \arg\sup_t f(t)$.

Estimates

A. Functions of single variable:

1. $\inf[f_1(x) + f_2(x)] \geq \inf f_1(x) + \inf f_2(x)$.
2. $\inf[f_1(x) f_2(x)] \geq \inf f_1(x) \inf f_2(x)$ if $f_1(x) > 0$, $f_2(x) > 0$.
3. $\inf \dfrac{f_1(x)}{f_2(x)} \geq \dfrac{\inf f_1(x)}{\sup f_2(x)}$, if $f_1(x) > 0$, $f_2(x) > 0$.

We gave above in 1–3 the sign $=$, if $\bar{x}_1 = \bar{x}_2$.

4. $\inf_{x(t)\in Q} \int_{t_1}^{t_2} f(t, x(t)) dt \geq \int_{t_1}^{t_2} \inf_x f(t, x) dt$.

B. Function of two variables:

1. 1. $\inf[f_1(x) + f_2(y)] = \inf_x f_1(x) + \inf_y f_2(y)$.
2. $\inf[f_1(x) f_2(y)] \geq \inf_x f_1(x) \cdot \inf_y f_2(y)$ if $f_1(x) \geq 0$, $f_2(y) \geq 0$.
3. $\inf_{x,y} \dfrac{f_1(x)}{f_2(y)} \geq \dfrac{\inf f_1(x)}{\sup f_2(y)}$, if $f_1(x) \geq 0$, $f_2(y) > 0$.
4. $\inf_{x,y} f(x, y) = \inf_x \inf_y f(x, y) = \inf_y f(\bar{x}(y), y)$.

2. Exercises for β- and γ- functions

Choosing β- function, find quasi-optimal solutions to precision 5%.

Indication: Find the lower estimate. Separate subset which contains points of global minimum and take quasi-Optimal solution from it.

Examples: Answers:

1. $I = x^4 + x^2 + 0.2x + 1$, $M = \{x: -0.2 \leq x \leq 0\}$, $I(0) = 1 \geq 0.99$.
2. $I = x^6 + x^2 + 0.2x + 1$, $M = \{x: -0.2 \leq x \leq 0\}$, $I(0) = 1 \geq 0.99$.
3. $I = x^8 + x^2 - 0.2x + 1$, $M = \{x: 0 \leq x \leq 0.2\}$, $I(0) = 1 \geq 0.99$.
4. $I = x^{2n} + x^2 - 0.2x + 1$, $M = \{x: 0 \leq x \leq 0.2\}$, $I(0) = 1 \geq 0.99$.
5. $I = |x|^m + x^2 - 0.4x + 1$, $M = \{x: -0.2 \leq x \leq 0\}$, $I(0) = 1 \geq 0.99$.
6. $I = |x|^m + 2x^2 - x + 3$, $M = \{x: -0.5 \leq x \leq 0\}$, $I(0) = 3 > 2\tfrac{7}{8}$.
7. $I = x^2 - 4x + 6 - 0.1\sqrt[3]{e^{-(x-1)^2}}$, $M_1 = \{x: 0 \leq x \leq 2\}$, $I(2) = 2 - 0.1e^{-2/3} \geq 1.9$.
   $M_2 = \{x: 1 \leq x \leq 3\}$.

8. $I = x^2 - 4x + 6 - \dfrac{0.1}{(x-1)^2 + 10}$, $M = \{x : 1 \le x \le 3\}$, $I(2) = 2 - \dfrac{1}{10} \ge 1.99$.

9. $I = x^2 - 2x + 5 - \dfrac{1}{x^2 - 4x + 14}$, $M = \{x : 1 \le x \le 2\}$, $I(1) = 4 - \dfrac{1}{11} \ge 3.9$.

10. $I = x^2 + 4x + 6 - \dfrac{0.1}{x^4 + 3x^3 + 3x^2 + 2}$, $M = \{x : -3 \le x \le -1\}$, $I(-2) = 1.95 \ge 1.9$.

11. $I = x^2 + 2x + 3 - \dfrac{0.1}{e^{x^2 - 2x + 1}}$, $M = \{x : -3 \le x \le 1\}$, $I(-1) = 2 - \dfrac{0.1}{e^2 + 1} \ge 1.9$.

12. $I = |x - 1|^3 + 5 - \dfrac{0.2}{(x-1)^2 + 10}$, $M = \{x = 1\}$, $I(1) = 4.98 \ge 4.98$.

13. $I = \sqrt{x^2 - 4x + 8} - \dfrac{0.1}{(x-1)^2 + 5}$, $M = \{x : 1 \le x \le 3\}$, $I(2) = 2 - \dfrac{0.1}{6} \ge 2 - \dfrac{0.1}{5}$.

14. $I = \sqrt[3]{x^2 - 4x + 12} - \dfrac{0.1}{\sqrt{x^2 - 2x + 5}}$, $M = \{x : 1 \le x \le 3\}$, $I(2) = 2 - \dfrac{0.1}{\sqrt{5}} \ge 2 - \dfrac{0.1}{2}$

15. $I = \sqrt{x^2 + 4x + 8} - \dfrac{0.1}{\sqrt[3]{x^4 + 3x^3 + 3x^2 + 2}}$, $M = \{x : -3 \le x \le -1\}$, $I(-2) = 1.95 \ge 1.9$

16. $I = |x - x_1|^n + c_1 - \dfrac{d}{|x - x_2|^m + c_2}$, $d > 0$, $c_2 > 0$, $n > 0$, $m > 0$,

    $M = \{x : |x - x_2| \le |x_1 - x_2|\}$, $I(x) \ge c_1 - \dfrac{d}{c_2}$.

17. $I = \sqrt[k_1]{|x - x_1|^n + c_1} - \dfrac{d}{\sqrt[k_2]{|x - x_2|^m + c_2}}$, $d > 0$, $c_1 > 0$, $c_2 > 0$, $n > 0$, $m > 0$, $k_1 > 0$, $k_2 > 0$,

    $M = \{x : |x - x_2| \le |x_1 - x_2|\}$, $I(x) \ge \sqrt[k_1]{c_1} - \dfrac{c_1}{\sqrt[k_2]{c_2}}$.

18. $I = |x(x - 2)| - \dfrac{0.1}{|x - 1| + 10}$, $M = \{x : 0 \le x \le 2\}$, $I(0) = 0 - \dfrac{0.1}{11} \ge -\dfrac{0.1}{10}$.

19. $I = |x(x - a)| - \dfrac{d}{|x - b| + c}$, $d > 0$, $c > 0$,

    $M_1 = \{x : |x - b| \le |a - b|\}$, $M_2 = \{x : 0 \le x \le 2b\}$, $I \ge -\dfrac{d}{c}$.

20. $I = \dfrac{|x|}{x} + |x|$. $M = \{x : x < 0\}$ $I(0) = -1 \ge -1$. *Indication*: $\beta = -[x]$. $I(0) = -1 \ge -1$.

21. $I = x^2 - 1.8x + 1 + \dfrac{1}{1+e^{\sin x}}$. Answer $M = \{x: 0.8 \le x \le 1\}$ $I(1+0) = 0.2 - 0 \ge 0.19$.

22. $I = x^2 - 0.2x + 10.1 + \dfrac{1}{10 + \lg|\cos x|}$, Ansfer $M = \{x: |\cos x| \ge 10.10\}$ $I(0) = 10.1 \ge 10.1$

23. $I = x^2 + xe^{4x} + 10$. Ans. $M = \{x: x \le 0\}$ $I(0) = 10 \ge 10 - \dfrac{1}{4e}$.

24. $I = \dfrac{e^x}{x} + x\ln^2 x$, $x \ge 0$. Ans. $M = \{x: 0 < x \le 1\}$ $I(0) = e \ge e$.

25. $I = x^6 + y^6 + 2x^2 - 4xy + 2y^2$. Ans. $M = \{x,y: x = y\}$ $I(0,0) = 0 \ge 0$.

26. $I = |x| - e^{-y^2} + x^2 - 2xy + y^2$. Ans. $M = \{x,y: x = y\}$ $I(0) = 1 \ge 1$.

27. $I = |x| + |y-1| + |z+1| + \dfrac{1}{e^{x^2+y^2+z^2}} + 6$. Ans. $M = \{x,y,z: x^2+y^2+z^2 \le 2\}$ $I(0,1,-1) = 6 + \dfrac{1}{e^2} \ge 6$.

28. Find the minimum from all integer solutions of function
$$I = (x-32)^2 + \dfrac{(\log_2 x - 5)(\log_2 x - 5.1)}{\lg x}.$$

Indication. Take as β the second member in I and consider the in the extended area $0 < x < \infty$. We find $M = \{x: 32 \le x \le 34.3\}$. Calculate I for $x = 32, 33, 34$ and select better.

Find the lower estimation by using the γ – function.

29. $I = \dfrac{x^2(x-2)^2}{2 - \sin x}$. Ans. $I(x) = 0$, $x_1^* = 0$, $x_2^* = 2$.

30. $I = (x-2)^2(1 + \lg^2 x)$. Ans. $I(x) \ge 0$, $x^* = e$.

31. $I = (|x| + |y|)e^{-(x^2+y^2)}$. Ans. $M = \{0,0\}$, $I(0,0) = 0 \ge 0$.

References to Chapter 1

1. Bolonkin A.A., A certain method of solving optimal problems, Processing Siberian Department of USSR Science Academia (Izvesiya Sibirsk. Otdel. Akadem.Nauk SSSR), 1970, No.8, issue 2, June, pp.86-92 (in Russian). Math. Reviews, v.45, #6163 (English).
2. Bolonkin A.A., New Methods of Optimization and their Application, Moscow, MVTU, 1972, 220 ps. (Russian). Болонкин А.А., **Новые методы оптимизации и их применение**. МВТУ им. Баумана, 1972г., 220 стр. (См. РГБ, Российская Государственная Библиотека, Ф-861-83/1809-6**).**
http://vixra.org/abs/1504.0011 **v4.**   http://viXra.org/abs/1501.0228, (v1, old) ,
http://viXra.org/abs/1502.0137 **v3;** http://viXra.org/abs/1502.0055 v2;
https://archive.org/details/BookOptimization3InRussianInWord20032415 v2,
https://archive.org/details/BookOptimization3InRussianInWord20032415_201502 v3,
https://archive.org/details/BookOptimizationInRussian **(old),**

http://www.twirpx.com/file/1592607/ 2 6 15 загрузил v2.
http://www.twirpx.com/file/1605604/?mode=submit   v3, загрузил 2 16 15
https://www.academia.edu/11054777/   v.4.

3. Болонкин А.А., **Новые методы оптимизации и их примеение в задачах динамики управляемых систем.** Автореферат диссертации на соискание ученой степени доктора технических наук. Москва, ЛПИ, 1971г., 28 стр.   http://viXra.org/abs/1503.0081, 3 11 15.
http://www.twirpx.com.
http://samlib.ru/editors/b/bolonkin_a_a/ , http://intellectualarchive.com/ .#1488.
https://independent.academia.edu/AlexanderBolonkin/Papers,

4. **Докторская диссертация** А.Болонкина: Новые метоы оптимизации и их применение в задачах динамики управляемых систем. ЛПИ 1969г.
https://drive.google.com/file/d/0BzlCj79-4Dz9YTJOUHVhR1FZUVE/view?usp=drive_web Dissertation Optimization 1-2 9 30 15.doc, http://viXra.org/abs/1511.0214 , http://viXra.org/abs/1509.0267 Part 1, http://vixra.org/abs/1509.0265 Part2,
 https://archive.org/details/NewMethodsOfOptimizationAndItsApplication.Part1inRussian
https://www.academia.edu/s/2a5a6f9321?source=link, http://www.twirpx.com,

5. Болонкин А. А., **Об одном методе решения оптимальных задач**. Известия СО Академии наук СССР, вып.2, № 8, июнь 1970 г. http://www.twirpx.com/file/1837179/ ,
http://viXra.org/abs/1512.0357 , http://vixra.org/pdf/1512.0357v1.pdf ,   https://www.academia.edu
, https://archive.org/download/ArticleMethodSolutionOfOptimalProblemsByBolonkin

6. **List #1 Bolonkin's publications in 1965-1972.**(in Russian).   https://archive.org/details/No1119651972

7. Bolonkin A.A., A New Approach to Finding a Global Optimum, Vol.1, 1991, The Bnai Zion Scientists Division, New York (English).

8. Bolonkin A.A., **Impulse solutions in optimization problems** 11 20 15
http://viXra.org/abs/1511.0189;   https://www.academia.edu/s/00538971c8 ,
 https://archive.org/details/ArticleImpulseSolutionsdoc200311115AfterJoseph ,
 GSJornal: http://gsjournal.net/Science-Journals/Research%20Papers-Astrophysics/Download/6259,
 www.IntellectualArchive.com, #1625.

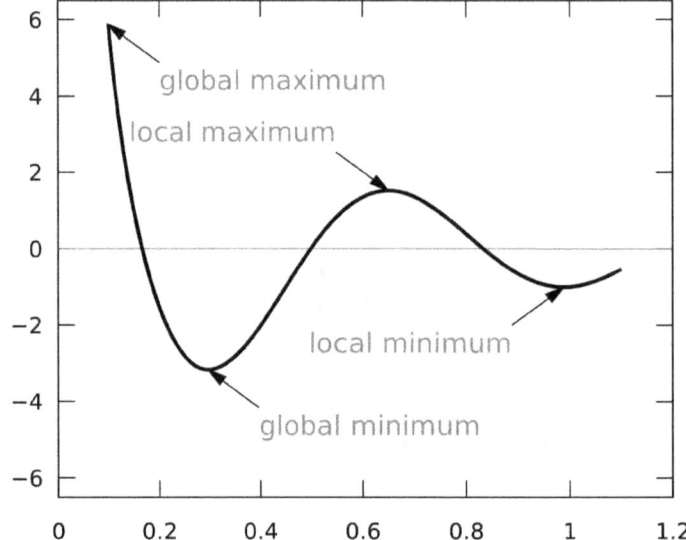

Chapter 2

Methods of α – functions. Estimations.

§1. α – functions over arbitrary set.

A. The special case of β-function is α-function. It is defined over set $Z = X \times Y$ and has the following properties:
1) There exist subset $K \subset Z$ with projection K on X, $pr_1 K = X^*$.
2) $\tilde{\alpha}(x, y) = 0$ on K.

Theorem 1.1. Assume $\tilde{\alpha}(x, y)$ is $\tilde{\alpha}$-function and exist the point of global minimum $x^* \in X^*$. Then the element $\bar{x}$ is point of the global minimum of object function $I(x)$ over set X^* if and only if there exist $\tilde{\alpha}(x, y)$ such that:

1) $J(\bar{x}, \bar{y}) = \inf[I(x) + \alpha(x,y)]$ $x, y \in Z$; 2) $\bar{x}, \bar{y} \in K$.

Proof: As $\bar{x}, \bar{y} \in K$, then $\alpha(\bar{x}, \bar{y}) = 0$ and

$$J(\bar{x}, \bar{y}) = \inf_Z [I(x) + \tilde{\alpha}(x,y)] = \inf_K [I(x) + \alpha(x,y)] = \inf_{X^*} I(x).$$

Q.E.D.

One may made vice versa. Define set $K_1 = \{x, y : \tilde{\alpha}(x, y) = 0, \; x \in X, \; y \in Y\}$. Find $X_1 = pr_1 K_1$. Then $\bar{x}$ is the point of minimum $I(x)$ over X_1, if $\bar{x}, \bar{y} \in K_1$.

The special case of $\tilde{\alpha}$-function is α-function defined over Z and such that $\alpha(x,y) = 0$ over X^* for all $y \in Y$.

The following theorem is important:

Theorem 1.2. Let us assume $\alpha(x,y) = 0$ over X^* for all $y \in Y$ and there exist $x^* \in X^*$. The element $\bar{x}$ will be the point of global minimum of objective function $I(x)$ over X^* if there exist function $\alpha(x,y)$ such that

1) $J(\bar{x}, y) = \inf[I(x) + \alpha(x,y)]$ $x, y \in Z$; 2) $\bar{x} \in X^*$. (1.1)

Proof: As $\bar{x} \in X^*$, then $\alpha(x, y) = 0$ and

$$J(\bar{x}, \bar{y}) = \inf_Z[I(x) + \alpha(x,y)] = \inf_X[I(x) + \alpha(\bar{x},y)] = \inf_{X^*} I(x).$$

Q.E.D.

If y is not constant, one can use it (the function $\alpha(x, y)$ from y) for getting $\bar{x} \in X^*$.

Theorem 1.3. $\tilde{\alpha}$ and α – functions exist and their number is infinite.

Theotem 1.4. (Estimate). If in (1,1) $\bar{x} \notin X^*$, we have a lower estimation of the objective function $I(x)$ on X^*:

$$J(\bar{x}(y), y) \leq I(x) \quad \text{for all} \quad y \in Y.$$

One can get this estimation from $\alpha(x,y) = 0$ on set X^* for all $y \in Y$ and **Principle of Extension**[1] [1], because $X^* \subseteq X$.

---

1) The Principle of Extension state: any extension of set, which you find on a minimum of functional, can only decrease on a minimum of an objective function (can only decrease value of a minimum). The Principle of Extention at first was published by A. Bolonkin "Principle of Extention and Yakoby codition of variable calculation", Report of Ukraine Science Academy #7, 1964 (ДАН УССР, №7, 1964)(in Ukraine).

The dependence J(x,y) from y one may use for improving of estimation. In particular, one can take α = α(x). Then from theorems 1.2, 1.3 one can get the following consequences:

<u>Consequence 1.</u> Assume α(x) = 0 on X^* and exist $x^* \in X^*$. Element $\bar{x}$ is point of a minimum of the objective function I(x) on X* if and only if the exist α(x) such, that

1) $J(\bar{x}) = \inf[I(x) + \alpha(x,y)]$ $x \in X$; 2) $\bar{x} \in X^*$. (1.1')

<u>Consequence 2</u>. If $\bar{x} \in X^*$, $\beta \equiv \alpha$ then $\inf_{X \times Y} J = \inf_{X^*} I$.

As far as α-function is the particular case β-function consequently the theorem 1.1 of Chapter 1 is right in this case.

Theorem 1.5. Assume $\bar{x}$ is point of global minimum of Problem 2:

$$J(\bar{x}) = \inf[I(x) + \alpha(x)], \quad x \in X.$$

Then: 1) The points of global minimum of Problem 1 are in the set

$M^* = M \cap X^*$, where $M = \{x : \alpha \geq \bar{\alpha}\}$;

2) Set $N^* = N \cap X^*$, where $N = \{x : J + I \leq \bar{J} + \bar{I}\}$, contain same or better solution that is in N the object function $I(x) \leq I(\bar{x})$;

3) Set $P^* P \cap X^*$, where $P = \{x : \alpha \leq \bar{\alpha}\}$ contains same or worse solutions (that is $I(x) \geq I(\bar{x})$ in P).

The same way for this case we can be formulated the Theorem 1.1

Since the set X^* is selected by equal $\alpha(x) = 0$ we get from Theorem 1.5 the consequences:

Consequence 3: If $\alpha(\bar{x}) > 0$, then $X^* \subseteq P$.

Consequence 4: If $\alpha(\bar{x}) < 0$, then $X^* \subseteq M$.

Consequence 5: If $\alpha(\bar{x}) = 0$, then $\bar{x} \in X^*$.

From Theorems 1.2 – 1.4 and Consequence 1 we get

Algorithm 4 :

We take the bounded of below functional (objective function) defined on X*Y, find minimal $\bar{x} = \bar{x}(y)$ of Problem 2: $\inf(I + \alpha)$, $x \in X$ or minimal in implicit form $\xi(\bar{x}, y) = 0$. We solve together the system equations (combining equations of α- function): $\xi(\bar{x}, y) = 0$, $\alpha(x, y) = 0$. Then value $\bar{x}$ - root pf this system is the absolute minimal of Problem1: $\inf(I + \alpha)$, $x \in X$.

Algorithm 4' (solution by choice of α-function).

We take the bounded of below functional α defined on X (or X*Y), solve the Problem 2: $\inf (I + \alpha)$, $x \in X$. If $\bar{x} \in X^*$, we get minimal of Problem 1, if $\bar{x} \notin X^*$, we get the estimation below $J(x) \leq I(x^*)$ of value of the objective function I(x) on set X^* and we get the sets M, N, P.

Comments:

1. If the admissible set X^* allocates by functional $F_i(x) = 0$, you can find the α - functional in form $\alpha = \lambda_i(x) F_i(x)$ (here *i* means sum), where $\lambda_i(x)$ are some function of x.

2. If the admissible set allocate by functional $\Phi_j(x) \leq 0$, you can find α – functional in form
$$\alpha = \omega_i(x)[\Phi_i(x) + |\Phi_i(x)|],$$
where $\omega_i(x)$ are some function of x, or in form
$$\alpha = \omega_i(x)\Phi_i(x),$$
where $\omega(x) \geq 0$ and it is fulfilled the condition $\omega_i(x)\Phi_i(x) \equiv 0$ on X^*.

3. Assume there is some α –functional and element $x \in X^*$ such $J(\bar{x}) = \inf[I(x) + \alpha(x)]$, $x \in X$.
Then any element $x_1 \in X^*$ and is satisfying the condition
$$J(x_1) = \inf[I(x) + \alpha(x)], \quad x \in X. \tag{1.1''}$$
is point of the absolute minimum the functional I(x) on X^* and any point of the absolute minimum the functional I(x) on X^* satisfy the condition (1.1").

This direct statement follows immediately from condition 1.

We proof the converse. Since the global minimal $x_1 \in X^*$, it means $\alpha(x_1) = 0$, then
$$I(x_1) = \inf_{X^*} I(x) = J(x_1) = J(\bar{x}) = \inf_X [I(x) + \alpha(x)].$$

Q.E.D.

Thus if it is exist one element which satisfy (1.1) then all rest minimal elements of Problem 1 must satisfy it. I illustrate the idea of α-functional the next sample.

Let us take some function f(x) definite on interval [a, b]. Digital values $n \in [a, b]$ are admissible for it. We want find the minimum of this function. The addition member (α –functional) do not change f(n) in points n, but deforms f(x) in gaps between n (see fig. 2.1).

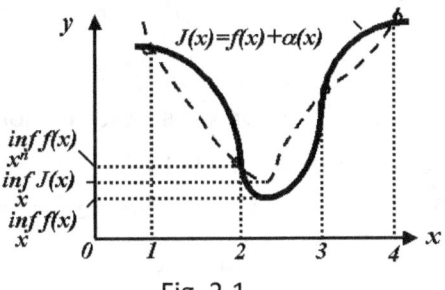

Fig. 2.1.

If α – functional is "good", then $\inf_{x\in[a,b]}[f(x)+\alpha(x)] > \inf_{x\in[a,b]} f(x)$. If in addition $\bar{x}=n$, then we get the minimum of Problem 1.

Remark: There are different ways to solve problems by the α-functional:

a) You can take the known function as α-functional.
b) You can take α-functional as unknown function and find it together with the point of minimum.
c) You can take α-functional as function α = α(x,y) where α is known function but y = y(x) is unknown function of x. You must find it together with the point of minimum.

Let us consider the example. We take as example the non-good the functional which is difficult to solve by conventional method.

Example 1.1. Find the minimum of function

$$I = \frac{4x^2+4\pi x+4.1+\pi^2}{4(x^2+\pi x+1)+\pi^2} \cdot \frac{\sin^5 x - \sin^4 x \cdot \cos x + \sin^2 x \cdot \cos^3 x}{(\sin x - \cos x)(\sin^3 x + \cos^3 x)} \quad \text{in } X^* = \{x=0.5\pi n: n=0,\pm 1,\pm 2,...\} \quad (1.2)$$

It is difficult to apply the known methods here because the functional is defined on digital set. The current methods offer only the calculation of all $x \in X^*$. But number of X^* equals infinity and calculation may be meaningless.

Let us to solve this example by the offered method. Take α(x) in form

$$\alpha = -\frac{4x^2+4\pi x+4.1+\pi^2}{4(x^2+\pi x+1)+\pi^2} \cdot \frac{0.5\sin 2x \cdot \cos x}{(\sin x - \cos x)(\sin^3 x + \cos^3 x)}.$$

You can see that α(x) = 0 in X^* because for x = 0.5πn n = 0, ±1, ±2,...,   sin 2x = sin πn = 0.

Let us to create the general functional

$$J = I+\alpha = -\frac{4x^2+4\pi x+4.1+\pi^2}{4(x^2+\pi x+1)+\pi^2} \cdot \frac{\sin^5 x - \sin^4 x \cdot \cos x + \sin^2 x \cdot \cos^3 x - 0.5\sin 2x \cdot \cos x}{(\sin x - \cos x)(\sin^3 x + \cos^3 x)}.$$

Here the variable x is uninterrupted and $-\infty < x < \infty$ (set X)

The additive α(x) allows to change the functional (1.2) to simple form

$$J = \frac{4x^2+4\pi x+4.1+\pi^2}{4(x^2+\pi x+1)+\pi^2} \cdot \frac{(\sin^2 x - \cos^2 x)(1-\sin x \cdot \cos x)\sin x}{(\sin^2 x - \cos^2 x)(\sin^2 x - \sin x \cdot \cos x + \cos^2 x)} = \left(\frac{0.1}{4+(2x+\pi)^2}+1\right)\sin x.$$

This general functional is simple. His minimum may be found the conventional method of theory the function one variable. Here $\bar{x} = -\pi/2$, $\bar{x} \in X^*$ for $\bar{n}=1$, $\bar{I} = -1.025$. Consequently, that is absolute minimum (and sole) of initial functional (1.2).

We can apply an analogical method for finding of minimum on x the next functional

$$I = \cos^2 \varphi + 0.5\cos 2x\cos 2\varphi - 2\cos x\cos\varphi\cos(x+\varphi)+0.5-0.1e^{-x^2}, \quad X^* = \{x=0.5\pi n: n=0,\pm 1,\pm 2,...\}.$$

Here φ is given, x is digital. Let us take $\alpha = -0.5\sin 2x\sin 2\varphi$. After this we can change our functional J = I + α to simple form: $J = -0.1e^{-x^2}+\sin^2 x$. The point of absolute minimum this task (Problem 2) is $\bar{x}=0$. This point is in allowable set X^* for $\bar{n} = 0$. That means $\bar{n} = 0$ is point of the absolute minimum od the initial Problem 1.

The reader can think: if the allowable numerical set is limited we can use the conventional Lagrange's method. Let us show: that is not correct.

Example 1.2. Find minimum of functional:
$$I = x^3 - 3x^2 + x \quad on \quad X^* = \{x = 0, x = 3\}. \tag{1.3}$$
Let us to write the Lagrange's function
$$F = x^3 - 3x^2 + 2x + \lambda_1 x + \lambda_2(x - 3),$$
where λ_1, λ_2 are LaGrange's factors. Find the first derivative
$$F' = 3x^2 - 6x + 2 + \lambda_1 + \lambda_2.$$
Substitute to here $x = 0$, $x = 3$ and write the equations $F'(0) = 0$, $F'(3) = 0$. We find from these equations λ_1, λ_2. Find the second deviation $F'' = 6x - 6$. When $x = 0$ the function $F''(0) = -6 < 0$. When $x = 3$ the function $F''(3) = 12 > 0$. Consequently $x = 0$ is the point of maximum, $x = 3$ is the point of minimum. Let us check up. Substitute $x = 0$ and $x = 3$ in (1.3). We find $I(0) = 0$, $I(3) = 6$.

We see the LaGrange's method gives the opposed result: it declare the point of minimum as the point of maximum, but the point of maximum as the point of minimum. In here it is violate one condition of LaGrange's method: The number of additional equations is more of number of variables. This example is shows: this violation for LaGrange's method is unacceptable.

Let us to solve this example by the offered method. Take the α(x) in form
$$\alpha = x(x-3)(2/3-x).$$
Then
$$J = I + \alpha = x^3 - 3x^2 + 2x + x(x-3)(2/3 - x), \quad J' = 4/3x = 0, \quad \bar{x} = 0 \in X^*, J'' = 4/3 > 0.$$

From Consequence 1 the point $\bar{x} = 0$ is absolute minimum of functional (1.3). That shows the method of α-functional has more application then the LaGrande's method.

Example 1.3. Find minimum of integral
$$I = \int_{-10^{-3}}^{a} (\ln \text{tg}\, t - 10^{-3}) dt \quad on \quad X^* = \{a = 10^{-3} \pi n : n = 1, 2, \ldots, 400\} \tag{1.4}$$

Here the interval of integration is discrete. The direct search is difficult because integral (1.4) cannot be presented by simple function and it not have of tabulations.

Let us to find α-functional in form: $\alpha = -10^{-6} \sin 10^3 a$. You see on X^* the function α(x) = 0. Further

$$J = I + \alpha = \int_{10^{-3}}^{a} (\ln \text{tg}\, t - 10^{-3}) dt - 10^{-6} \sin 10^3 a,$$
$$J'_a = \ln \text{tg}\, a - 10^{-3} - 10^{-3} \cos 10^3 a = 0, \quad \bar{x} = \pi/4 \in X^* \quad for \quad \bar{n} = 250, \tag{1.5}$$
$$J'' = \frac{2}{\sin 2a} + \sin 10^3 a.$$

As $10^{-3} < x < 0.4\pi$, then $J'' > 0$ into this interval. That means the root is single and $\bar{n} = 250$ is point of the absolute minimum.

Analogically we find the minimum of other integral which cannot be presented in simple functions

$$I = -\int_0^a [\sin(t^3) + 10^{-5}\sqrt{\pi}]dt \quad on \quad X^* = \{a = 10^{-3}\sqrt{\pi n} : n = 0, 1, ..., 1.5 \cdot 10^3\}. \quad (1.6)$$

Here is $\alpha = 10^{-3}\sin 10^{-8}\sin 10^3\sqrt{\pi}a$; $\bar{n} = 1000$.

Example 1.4. Find the minimum of integral

$$I = \int_{\pi/2}^{\pi} \left(\frac{\cos at}{t} + 20a^3\right)dt \quad on \quad X^* = \{a = 10^{-3}n : n = 0, \pm 1, \pm 2, ...\}. \quad (1.7)$$

Here the under integral function is discrete. The integral from this function cannot be presented as elementary functions.

Let us take $\alpha = 10^{-3}\sin^2 10^3 \pi a$, $J = I + \alpha$. Then

$$J'_a = I'_a + \alpha'_a = \int_{\pi/2}^{\pi}(-\sin at + 40a)dt + 2 \cdot 10^{-4}\pi\sin 2 \cdot 10^3 \pi a =$$
$$= -\frac{2}{a}\sin\frac{3}{4}\pi a \cdot \sin\frac{\pi}{4}a + 20\pi a + 10^{-4}\pi\sin 2 \cdot 10^3 \pi a \quad (1.8)$$

This derivative not exist for $\bar{a} = 0 \in X^*$.

For $a \geq 0$, $J' > 0$; for $a < 0$, $J' < 0$; (or $J'' > 0$ for $\forall a \neq 0$).

Consequently $\bar{n} = 0$ is point of absolute minimum.

B) Consider the case when the point of optimum $x^* \in X^*$ not exist, but exist the sequence such that $\lim_{n\to\infty} I(x_n) = m$. This sequence is named the minimizing sequence (see §5 of Ch.1).

Similarly point A we can show that consequence 1 can be generalized in this case.

Consequence 1'. Let us $\alpha(x) = 0$ only on X^*, For minimizing sequence $\{x_n\} \subset X^*$ is necessary and sufficient the existing of function $\alpha(x)$ such that

$$\lim_{n\to\infty}[I(\alpha_n) + \alpha(x_n)] = \inf[I(x) + \alpha(x)], \quad x \in X. \quad (1.9)$$

We can generalize remark 3 of item 1 in this case: If exist α function and one sequence $\{x_n\} \subset X^*$ which satisfy (1.9), then the any sequence $\{x_n\} \subset X^*$ which satisfy (1.9) is the minimizing sequence. And on the contrary any the minimizing sequence satisfy the condition (1.9).

2. α – function in Banach space.

Let us to apply Theorem 1.2 to optimal problem is described in Banach space by equation

$$\frac{dx}{dt} = f(x,u), \quad t_1 \leq t \leq t_2, \quad x(t_1) = x_1, \quad x(t_2) = x_2, \quad (1.10)$$

where x, $f(x,u)$ – element complete linear normed space X_1 and X_2 respectively and $X_2 \subset X_1$, $t \in [t_1, t_2] = T$ is segment of real axis.

Let us name the permissible control the measurable limited function (in term [2], p.85) with value $u \in U$, where U is set in arbitrary topological space. In particular the set U may be metric, closed and limited. Let us assume that for any control $u(t)$ the equation (1.10) has single solution $x(t)$ with $x(t) \in X_1$ for almost all $t \in [t_1, t_2]$, where $x(t)$ is continuous almost everywhere differentiable on function on $t \in [t_1, t_2]$.

Operator $f(x,u)$ is defined on the direct product $X \times U$. One is continuous and bounded. Boundary conditions are given t_1, t_2, $x(t_1) = x_1$, $x(t_2) = x_2$.

State the problem: Find the admissible control which transfers the system from given initial state in given final state with function

$$I = \int_{t_1}^{t_2} f_0(x, y) dt \qquad (1.11)$$

has a minimum.

Let us the set of the measurable functions $u(t)$ is denoted V: set of the continuous, almost everywhere differentiable on (t_1, t_2) the functions $x(t)$ is denoted D. Set of couple $x(t)$, $u(t)$ having named over properties and almost all satisfied the equation (1.10), we name *admissible* and denote Q. It is obvious $Q \subset D \times V$.

Assume $\psi = \psi(t, x)$ is the some unequivocal continuous differential **function defined on $X \times T$. We name it the *characteristic function*. We will find the α – function in form**

$$\alpha = \int_{t_1}^{t_2} \psi_x * [\dot{x} - f(x,u)] dx \qquad (1.12)$$

Here $\psi_x = \dfrac{\partial \psi}{\partial x}$ is particular deviation of Freshe. One is linear function. The * is sign of composition. Obvious that request of α-function is performed.

Compose the generalized function $I = J + \alpha$ and produce the function $\dot{\psi} = \psi_x \dot{x} + \psi_t$ we get

$$J = \psi[t_2, x(t_2)] - \psi[t_1, x(t_1)] + \int_{t_1}^{t_2} (f_0 - \psi_t - \psi_x \circ f) dt = \psi_2 - \psi_1 + \int B dt \quad, \quad (1.13)$$

where $B = f_0 - \psi_t - \psi_x \circ f$. Because the set Q is different from the set $D \times V$ only that couple $x(t)$, $u(t)$ satisfy almost every where (1.10). For α-function in form (1.12) with according of Theorem 1.2 we can the initial Problem 1 (find the minimum (1.11) on Q) replace the Problem 2 – find minimum (1.13) on the broader set $D \times V$. In this set the $x(t)$, $u(t)$ not bind the equation (1.10). So we have

$$\overline{J} = \psi_2 - \psi_1 + \inf_{x(t) \in D, u(t) \in V} \int_{t_1}^{t_2} B(t, x, u) dt \,. \qquad (1.14)$$

Theorem 1.6. If function $\overline{u}(t)$ getting from solution of problem $\inf_{x(t) \in D, u(t) \in V} \int_{t_1}^{t_2} B dt$ is $\overline{u}(t) \in V$, that it is same almost everywhere the function getting from solution the problem $\inf_{\substack{x(t) \in D, \\ u(t) \in V}} \int_{t_1}^{t_2} B dt$ and

$$\inf_{x(t) \in D, u(t) \in V} \int_{t_1}^{t_2} B dt = \inf_{x(t) \in D} \int_{t_1}^{t_2} \inf_{u \in V} B dt \qquad (1.15)$$

Proof. Assume the contrary: $B(u^*) \neq \inf_{u \in U} B(u)$ on subset of interval $[t_1, t_2]$ with measure not equal zero. In this case $B(u^*) > B(\bar{u})$ i.e. $\int_{t_1}^{t_2} B(u^*) dt > \int_{t_1}^{t_2} B(\bar{u}) dt$ on the subset. This contrad**ict: the function** $u^*(t)$ made the minimum for integral $\int_{t_1}^{t_2} B dt$,

From requirement (1.14) and Theorem 1.6 we have

$$\bar{J} = \psi_2 - \psi_1 + \inf_{x(t) \in D} \int_{t_1}^{t_2} \inf_{u \in V} B(t, x, u) dt \tag{1.16}$$

If function $\alpha[x(t), u(t)]$ is such that absolute minimum of Problem (1.16): $\dot{x}(t)$, $\bar{u}(t) \in Q$, then αα according to Theorem 1.1 functions $\bar{x}(t), \bar{u}(t)$ are absolute minimum of the initial Problem.

So, we proofed

<u>Theorem 1.7</u>. To couple function were the absolute minimum the function *I*, it is sufficient the existing the characteristic function $\psi(t, x)$ such that

1) $B(t, x, \bar{u}) = \inf_{u \in U} B(t, x, u)$; 2) $\int_{t_1}^{t_2} B(t, x, \bar{u}) dt = \inf_{x(t) \in D} \int_{t_1}^{t_2} B(t, x, \bar{u}) dt$; 3) $\overset{r}{x}(t), \bar{u}(t) \in Q$; (1.17)

In particular, if take $\psi = p(t) \circ h$, where p(t) is linear function $h \in X_1$, then from item 1 and stationary condition item 2 [1.17] we get

$$H(t, x, \bar{u}) = \sup_{u \in U} \sup H(t, x, u), \quad \dot{p}(x) = -\frac{\partial H}{\partial x}, \tag{1.18}$$

where $H = p(t) \circ f(x, u) - f_0(x, u)$. Assumed $\partial H / \partial x$ is Frechet derivative, which is continuous. As we see the necessary condition of Problem 2 following from (1.17) is same the necessary condition of Pontriagin principe of maximum generalized in Banach spaces.

3. Design of α-function for allowable subset of two function connected by logical conditions

Assume two functions $F_1(x)$ and $F_2(x)$ are refinished on the set X. Allowable are only points $x \in X$ and functions F_1 and F_2 which are connected the logical conditions. Assume $F_1(x) = 0$ is "true" and $F_2(x) \neq 0$ is "false". The five main logical connections (↔, y, v, ∧, ~) (↔, y, ∨, ∧, ~) are presented in next tables:

F_1	F_2	$F_1 \leftrightarrow F_2$
t	t	t
t	f	f
f	t	f
f	f	t

Double implication

F_1	F_2	F_1 y F_2

t	t	f
t	f	t
f	t	t
f	f	f

disjunction in the exclusive sense

F_1	F_2	$F_1 \lor F_2$
t	t	t
t	f	t
f	t	t
f	f	f

disjunction in the sense of a non-exclusive

F_1	F_2	$F_1 \land F_2$
t	t	t
t	f	f
f	t	f
f	f	t

Conjunction

F	p
t	f
f	t

Denial

We will use the symbol:

$$\text{sign } F = 1 \quad \text{if} \quad F > 0,$$
$$\text{sign } F = 0 \quad \text{if} \quad F = 0,$$
$$\text{sign } F = -1 \quad \text{if} \quad F < 0,$$

In this case the α-function we can search in form:

1) $X^* = \{x : F_1(x) \leftrightarrow F_2(x)\}$, $\quad \alpha = (p_1 F_1 + p_2 F_2)[1- |\operatorname{sign}(F_1 F_2)|]$,
2) $X^* = \{x : F_1(x) \, y \, F_2(x)\}$, $\quad \alpha = p_1 F_1 F_2 + p_2[1- |\operatorname{sign}(F_1^2 + F_2^2)|]$,
3) $X^* = \{x : F_1(x) \vee F_2(x)\}$, $\quad \alpha = p F_1 F_2$,
4) $X^* = \{x : F_1(x) \wedge F_2(x)\}$, $\quad \alpha = p_1 F_1 + p_2 F_2$,
5) $X^* = \{x : F_1(x) \sim F_2(x)\}$, $\quad \alpha = (p[1- |\operatorname{sign} F|]$,

Here p, p_1, p_2 are some function x.

It is using these five connections we can create all other complex logic statements.

§2. The general principle of reciprocity the optimization problems

Let us suppose we want to solve the optimal problem Ch.1 §4 p.4 :
$$I = f_0(x), \quad f_i(x) = 0, \quad i = 1,2,\ldots,m, \tag{2.1}$$
Design general function in form
$$J = \sum_{i=0}^{i=n} \lambda_i(x,y) f_i(x), \tag{2.2}$$
where $\lambda_i(x,y)$ arbitrary functions of x, y.
Assume $\bar{x}(y)$ is absolute minimum (2.2) on X.

The general principle of reciprocity the optimization problems

1. For any $y \in Y$ the point of an absolute minimum of the function J (2.2) is the point of the absolute minimum any function
$$\lambda_j(x,y) f_j(x), \quad j = 0,1,\ldots,m \quad \text{(no sum for } j\text{)}, \tag{2.3}$$

for limits in form
$$\lambda_i(x,y) = \lambda_i(\bar{x}(y),y) f_i(\bar{x}(y)), \quad i = 0,1,\ldots,m, \quad i \neq j, \quad \text{(no sum for } i\text{)}. \tag{2.4}$$

Any numbers of equality (2.4) you can change by non-equalities
$$\lambda_i(x,y) \leq \lambda_i(\bar{x}(y),y) f_i(\bar{x}(y)). \tag{2.5}$$

2. For any $y \in Y$ the point of the absolute minimum of the function J (2.2) is point of the absolute minimum any sum the functions
$$\sum_j \lambda_j(x,y) f_j(x) \tag{2.3}'$$

for restrictions absent in sum (2.3)
$$\lambda_i(x,y) = \lambda_i(\bar{x}(y),y) f_i(\bar{x}(y)), \quad i = 0,1,\ldots,m, \quad i \neq j, \quad \text{(no sum for } i\text{)}. \tag{2.4}'$$

Any numbers of equality (2.4)' you can change by non-equalities (2.5).

Proof.

1) For any function (2.3) for conditions (2.4) the Theorem 1.2 is made. The point $\bar{x}(y)$ is point of its absolute minimum. As every function reaches the global minimum, obvious, the change equality (2.4) by restrictions (2.5) not influence to minimum. The point 2 is proofed similarly. Principle is proved.

Consequence 1.

Magnitude $J(\bar{x}(y), y)$ is the lower estimation of any function from (2.3), (2.3)' if part or all equalities (2.4), (2.4)' change equalities in form

$$\lambda_i(x, y) f_i(x) = 0 \qquad (2.6)$$

Consequence 2. In case corresponded (2.6) the absolute minimum of any functions (2.3) are located in set

$$M_j(y) = \{x : \sum_{\substack{i=1 \\ i \neq j}}^{m} \lambda_i(x, y) f_i(x) \geq \sum_{\substack{i=1 \\ i \neq j}}^{m} \lambda_i((\bar{x}(y), y) f_i(\bar{x}(y)) \} \qquad (2.7)$$

Consequence 3. If possible the solution of Problem (2.1) by Algorithm 4, there are y such that

$$\lambda_i((\bar{x}(y), y) f_i(\bar{x}(y)) \leq 0 \quad \text{(no sum for } i\text{)} \qquad (2.8)$$

From the existence of solutions (2.1) follows that $f_i(x) = 0$. So $\bar{\lambda}_i \bar{f}_i$ is minimum, than (2.8) is obvious.

§3. Applications α-function to well-known Problems of optimization

1. *Problem the searching of conditional extreme the function of the limited number variables.*

It is given

$$I = f_0(x), \quad f_i(x) = 0, \quad i = 1, 2, \ldots, m < n \qquad (3.1)$$

Here x is n-dimensional vector given in some numerical open region of n-dimensional space X^*.

Let us take the α-function in form

$$\alpha = p_i(x) f_i(x), \quad i = 1, 2, \ldots, m \qquad (3.2)$$

(repeated indexes means summarization). Here $p_i(x)$ are functions x, given on X:

$$X^* = \{x : \sum_{i=1}^{m} |f_i(x)| = 0\}, X^* = X.$$

Let us to design generalized functional $J(x) = f_0(x) + \alpha(x)$ take some $p_i(x)$ and sole the problem $\inf J(x)$, $x \in X$. From this solution the Problem 2, according Theorems §1, we can get the following information about Problem 1:

1) If $\bar{x} \in X^*$, than $\bar{x}$ is absolute minimum of Problem 1 (consequence 1, §1).
2) If $\bar{x} \notin X^*$, then:
   a) $J(\bar{x})$ is the lower estimation of function $f_0(x)$ on X^* (Theorem 1.4).
   b) For $\alpha(\bar{x}) > 0$ x^* is located in set $P = \{x : \alpha(x) \leq \alpha(\bar{x})\}$ (consequence 3, §1).
   c) For $\alpha(\bar{x}) < 0$ x^* is located in set $M = \{x : \alpha(x) \geq \alpha(\bar{x})\}$ (consequence 4, §1).

d) Set $N^* = N \cap X^*$ where $N = \{x : 2f_0 + \alpha \leq 2\bar{f}_0 + \bar{\alpha}\}$ contains the equal or worse solutions (Theorem 1.5).

As we see even if $\bar{x} \notin X^*$ our computation is useful. We received the lower estimation and narrow the region for searching of the optimal solution. Take row of α_i we can get the solution one of the Problems *a, b, c, d* or facilitate the solution of Problem *a* (see Ch. 1, §1).

Look your attention: the offered method does not require continuity and differentiability of the functions $f_0(x)$, $f_i(x)$ in contrast to the classical method of Lagrange multipliers. The method can be applied to non analitical function, for example, to the functions definished on the discret set and extremal problems of the combinatorics (see Ch. 10).

2. Application the Theorems §1 to optimal problems described the conventional differential equations.

Assume the moving of object is described by system of the differential equations

$$\dot{x}_i = f_i(t, x, u), \quad i = 1, 2, \ldots, n, \quad t \in T = [t_1, t_2], \tag{3.3}$$

where $x(t)$ – n-dimensional continuous piecewise differentiable function, $x \in G(t)$; $u(t)$ – r-dimensional functions continuous everywhere on T, except limited number of points where one can have discontinuity of the first kind $u \in U(t)$. Boundary values t_1, t_2 are given, $x(t_1)$, $x(t_2) \in R$.

Optimal function is

$$I = F(x_1, x_2) + \int_{t_1}^{t_2} f_0(t, x, u) dt, \quad x_1 = x(t_1), \quad x_2 = x(t_2). \tag{3.4}$$

Functions $F(x_1, x_2)$, $f_i(x, u, t)$, $i = 0, 1, \ldots, n$ are continuous, $F(x_1, x_2) > -\infty$. Set of the continuous almost everywhere differentiable functions $x(t)$ with $x \in G(t)$ we designate D. Set of the piecewise continuous (they can have the discontinuity of the first kind) functions u(t) such that $u \in U(t)$ we designate V. Couple x(t), u(t) have named over properties and almost everywhere satisfy the equations (3.3) we name allowable and designate Q, $Q \subset D \times V$.

Enter in our research n single-valued functions $\lambda_i(t.x)$ $i = 1, 2, \ldots, n$ which are continuous and have continuous derivatives on T×G. Let us to take the α-function in form

$$\alpha = \int_{t_1}^{t_2} \lambda_i(t, x)[\dot{x}_i - f_i(t, x, u)] dt \tag{3.5}$$

It is obvious $\alpha = 0$ on Q. Let us design the general function $J = I + \alpha$, integer the term $\lambda_i \dot{x}_i$ by part and exclude $\dot{x}_i$ by (3.3). We get

$$J = F + \lambda_i x_i \Big|_{t_1}^{t_2} + \int_{t_1}^{t_2} [f_0 - (x_j \frac{\partial \lambda_j}{\partial x_i} + \lambda_i) f_i - x_i \frac{\partial \lambda_i}{\partial t}] dt \tag{3.6}$$

Designate

$$a = F + \lambda_i x_i \big|_{t_1}^{t_2}, \quad B = f_0 - (x_j \frac{\partial \lambda_j}{\partial x_i} + \lambda_i) f_i - x_i \frac{\partial \lambda_i}{\partial t}$$

Apply to (3.6) Consequence 1 §1. Here the Q is X* and DxV is X (see Consequence 1 §1). Since now the couple of functions x(t), u(t) from DxV (having ends in R for condition $\bar{x}(t) \in D, \bar{u}(t) \in V, x_1 = x(t_1), x_2 = x(t_2)$) are not connected by the equations (3.3) we can write

$$\inf_{D \times V}(A + \int_{t_1}^{t_2} B dt) = \inf_{x_1, x_2 \in R} A + \int_{t_1}^{t_2} \inf_{x \in G, u \in U} B dt$$

and final

$$\bar{J} = \inf_{x_1, x_2 \in R} A + \int_{t_1}^{t_2} \inf_{x \in G, u \in U} B dt \tag{3.7}$$

So we proofed the **Theorem 3.1**:
The couple vector-function $\bar{x}(t), \bar{u}(t)$ will be point of absolute minimum of function (3.4) if it is exist n differentiable $\lambda_i(t,x)$ such that:

$$1) \bar{B} = \inf_{x \in G, u \in U} B, \quad 2) \bar{A} = \inf_{x_1, x_2 \in R} A > -\infty, \quad 3) \bar{x}, \bar{u} \in Q \tag{3.8}$$

Note: That is sufficient condition only. That cannot be a necessary condition because we don't know advance about an existence of $\lambda(t,x)$.

From (3.8) it is follow: if we find at least one solution of an equation in particular derivations having n-unknown functions $\lambda_i(t,x)$:

$$\inf_{u \in U}[f_0 - (x_j \frac{\partial \lambda_j}{\partial x_i} + \lambda_i) - x_i \frac{\partial \lambda_i}{\partial t}] = 0, \tag{3.9}$$

for boundary condition A = const, than points 1, 2 of the Theorem 3.1 will be executed. Any unsuccessful $\lambda_i(t,x)$ (if $\bar{x}(t), \bar{u}(t) \notin Q$) with according Theorem 1.4 gives the lower estimation of the global minimum.

Assume, for example, $x_n \neq 0$ *. Substitute them in (3.7), we get the result published in work [3]**, (condition Bellman-Pikone):

$$\bar{J} = \inf_{x_1 \in G_1, x_2 \in G_2} \Phi - \int_{x_1}^{x_2} \sup_{x \in G, u \in U} R(t,x,u) dt, \tag{3.10}$$

Here $\Phi = F + \varphi\big|_{t_1}^{t_2}, R = \varphi_t + \varphi_{x_i} f_i - f_0 = -B.$

-----

* This limitation is not important becouse any $x_i \neq 0$ in $[t_1, t_2]$.
** Note: in given method (in difference from known method) not request a priory assamption about existing the single potensial function $\varphi(t,x)$ such that $\varphi_{xi} = \lambda_i$.

Sometimes it is more comfortable take function $\varphi(t,x)$ or in other terms (see [4]) $\psi(t,x)$. Then A, B are written:

$$A = F + \psi_2 - \psi_1, \quad B = f_0 - \psi_{x_i} f_i - \psi_t, \tag{3.11}$$

And Theorema 3.1 is same the conventional condition [2]-[3].

Function α for given task we can define also the next way. Take some function $\psi(t,x)$. Than

$$\alpha = \int_{t_1}^{t_2} \psi_{x_i} [\dot{x}_i - f_i(t,x,u)]dt$$

Integrate but parts the first member we get

$$\alpha = \psi \big|_1^2 - \int_{t_1}^{t_2} (\psi_{x_i} f_i + \psi_t) dt$$

Note: 1. Theorema 3.1 is corrected and in notations (3.8) part 1:

$$\int_{t_1}^{t_2} B dt = \inf_{x(t) \in B} \int_{t_1}^{t_2} \inf_{u \in U} B dt.$$

This form is offered in [4]. Difference between these forms is important in consideration the second variation, conditions in angle points and in some other cases. Let us take the last corrected form of V. Krotov optimization [8] (problem of speed):

Example 3.1. Find minimum t_2 in task:

$$I = \int_{t_1}^{t_2} dt, \quad \dot{x} = u, \quad |u| = 1, \quad x(0) = 1, \quad x(t_2) = 0.$$

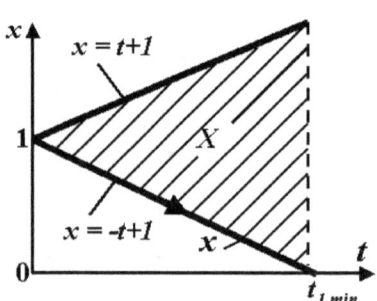

Fig.2.2.

If we take $\varphi = 0$, we get $R = -1$. Consequently $\sup_{x,u} R$ is reached in ANY curve, for example, $u = -0.01$ ($I = 100$).

In case whem *min* forward integral for $\psi = 0$ we have

$$\inf_{x(t) \in D} \int_{t_1}^{t_2} dt = \inf_{x(t) \in B} t_2[x(t)].$$

Since the set all serves with bounded derivative $|\dot{x}| \leq 1$ for $x(0) = 1$ located between lines $x = t - 1$, $x = -t + 1$ (Fig. 2.2), we get $\bar{x} = 1 - t$, $\bar{u} = -1$ and $I = t_{1,\min} = 1$.

Notes: 1. As set B we can take a set $\{x(t)\}$ with bounded derivative $\dot{x}_i \in \dot{X}_i = \{f_i(t,x,u) : u \in U\}$. This narrowing can help in finding of optimal solution.

2. Note 3 §1 in given case has the following view: If exist the function $\psi(t,x)$ and at list one allowable couple $\bar{x}(t), \bar{u}(t)$, satisfying (3.8). That any other couple satisfying (3.8) is minimum of problem 1 and any allowable minimum the problem 1 satisfy point 1, 2 (3.8).

3. If t_1, t_2 are not fixed, we can show that point 1, 2 (3.8) are:

$$1)\ \overline{B} = \inf_{x \in G,\ u \in U} B = 0, \quad 2)\ \overline{A} = \inf_{t_1, t_2, x_1, x_2 \in R} A > -\infty$$

We can satisfy the condition $\inf B = 0$, if we take $\psi = \varphi(t,x) + y_{n+1}$ and

$$\&_{n+1} = f_0 - \varphi_{x_i} f_i - \varphi_t.$$

4) Theorem 3.1 is particular case of more common theorem 2.1 considered in Chapter III.
Assume we take some $\lambda_i(t,x)$ (or $\psi(t,x)$).

Theorem 3.2. Assume $F = 0$ and solved the problem $\inf_{x,u} B$. That:

1) Set $N = \{t,x,u : B + f_0 \leq \overline{B} + \bar{f}_0, t \in T\}$ contains same and better solutions of Problem 1;

2) Set $P = \{t,x,u : B - f_0 \leq \overline{B} - \bar{f}_0, t \in T\}$ contains same and worse solutions of Problem 1.

Proof: 1) Deduct $B \geq \overline{B}$ from inequality $B + f_0 \leq \overline{B} + \bar{f}_0$. We get $f_0 \leq \bar{f}_0$ on T, i.e. $\int_T f_0 dt \leq \int_T \bar{f}_0 dt$.

2) Deduct $B \geq \overline{B}$ from inequality $B - f_0 \leq \overline{B} - \bar{f}_0$. We get $-f_0 \leq -\bar{f}_0$ on T, i.e. $\int_T f_0 dt \geq \int_T \bar{f}_0 dt$. Theorem is prooved (QED).

Let us take instead function (3.4) simpler function $\int_T B_1(t,x,u)dt$ (here B_1 is given function). These is

Theorem 3.3. Assume $F = 0$ and solved the problem $\overline{J}_1 = \inf \int_T B_1(t,x,u)dt$ on Q. Then:

1) Set $N = \{t,x,u : B_1 + f_0 \leq \overline{B}_1 + \bar{f}_0, t \in T\}$ contains the same and better solutions of Problem 1;

2) Set $P = \{t,x,u : B_1 - f_0 \leq \overline{B}_1 - \bar{f}_0, t \in T\}$ contains the same and worse solutions of Problem 1.

Proof: 1) From N we have the inequality $\int_T (f_0 + B_1)dt \leq \int_T (\overline{B}_1 + \bar{f}_0)dt$. Deduct from this inequality the inequality $\int_T B_1 dt \geq \int_T \overline{B}_1 dt$. We get $\int_T f_0 dt \leq \int_T \bar{f}_0 dt$.

2) From P we have the inequality $\int_T (B_1 - f_0)dt \leq \int (\overline{B}_1 - \bar{f}_0)dt$. Deduct $\int_T B_1 dt \geq \int \overline{B}_1 dt$ from this inequality. We get $\int_T f_0 dt \geq \int_T \bar{f}_0 dt$. Theorem is prooved (QED).

Consequence. If set P cover the set $T \times G \times U$ (or reachability set) and $\bar{x}, \bar{u} \in Q$, then $\bar{x}, \bar{u}$ are absolute minimum of Problem 1.

Note. Delete part equation (3.1) or (3.3)* [*in case (3.2) x_i correspoded deleted equations became the control in the rest equations]. Then getten solution is the low estimation of initial Problem as it is follow from principle of expension [5]: $I(x) \geq I(\bar{x})$ and $I(x,u) \geq I(\bar{x},\bar{u})$, where $\bar{x}(t), \bar{u}(t)$ are absolute minimum "truncated" task.

When right parts of equations (3.3), (3.4) do not depent clearly from x(t), we can stand out not only set N,P but the set M. It is correct the following theorema

Theorem 3.4. Assume $F \geq 0$, ends x(t) is free, the right parts of equations (3.3), (3.4) depent only from t, u, i.e.: $f_i = f_i(t,u)$ i = 0,1,...,n. and solved task $\inf_{x,u} B_1(t,u)$. Than:

1) Set $M = \{t, u : B_1 - f_0 \geq \bar{B}_1 - \bar{f}_0, t \in T\}$ contains the absolute minimum of Problem 1;

2) Set $N = \{t, u : B_1 + f_0 \geq \bar{B}_1 + \bar{f}_0, t \in T\}$ contains the same and better solutions of Problem 1;

3) Set $P = \{t, x, u : B_1 - f_0 \leq \bar{B}_1 - \bar{f}_0, t \in T\}$ contains the same and worse solutions of Problem1.

Proof for sets N, P full equally with the proof of Theorem 3.2. Proof for M follows from discontinuity u(t) and depends the right parts of equation only from u.

3. Task the dynamic programming of Bellman

Assume there is physical system S. The control of this system separated in *m* steps. On every *i* step we have the control U_i. Using this control we transver our system from allowable stand S_{i-1} getted in (*I* - 1) step in new allowable stand $S_i = S_i(S_{i-1}, Ui)$. This transwer is bounded by some conditions. The purpose is minimum function

$$W = \sum_{k=1}^{n} w_k$$

Let us to biuld the common function

$$J_i = W_i + \alpha, \text{ where } W_i = \sum_{k=1}^{n} w_k, i = 1,2,...,m$$

In this case we can change the task of the conditional minimum inf W_i in the task of direct minimum $\inf_V J_i$. If the limitations are absent or they allow the select U_k in every step to make with associated conditions, then from $\alpha = 0$ in the admisseble elements we get the Bellman equation [2].

$$\bar{W}_i(S_{i-1}) = \min_{U_i}\{W_i(S_{i-1}, U_i)\}, \quad i = 1,2,...,m.$$

3. Application α-function for solution the problems with distributed parameters

Let us consider about absolute minimum the Problem with distributed parameners

$$I(x,u) = \int_P f_0(t,x,u)dt + F(x(\tau)) \quad (3.12)$$

where $t = (t_1, t_2,...,t_m)$, $x = (x_1, x_2,...,x_n)$, $u = (u_1, u_2,...,u_r)$ are elements of vector space T, X, U * respectively. P is closed area in space T, bounded continuous piecewise smooth, fixed hypersurface S. On S the t = τ. P* is internal part this area, functions $x_i(t)$ on P are absolute-continious, $u_a(t)$ are measurable on P and have velues from area U, which can be closed and bounded.

Functions *x(t), u(t)* satisfy almost everywhere the system $n \cdot m$ indepedend differensial equetions with particulal deviations

$$\frac{\partial x_i}{\partial t_j} = f^i_j(t,x,u), \quad i=1,2,\ldots,n; \quad j=1,2,\ldots,m \tag{3.13}$$

Funsions f^i_j, f_0 are continiouly together with its particular derivitives the first order. The function *x(t), u(t)* we name allowable if they satisfy the named above conditions (set Q).

<u>Statement og Problem:</u> Find couple function *u(t), x(t)*, which give the function *I* (3.12) the minimal value.
Add to system (3.13) the integrability condision:

$$\varphi^\gamma = \frac{\partial f^i_j}{\partial t_k} - \frac{\partial f^i_{ki}}{\partial t_j} = 0, \quad i=1,2,\ldots n; \quad j,k=1,2,\ldots,m; \quad k>j. \tag{3.14}$$

Not difficult to calculate, that number of difficalt equation (3.14) may be $0.5(m-1)mn$, i.e. $\gamma=1,2,\ldots,0.5(m-1)mn$ (number of combinations $C^2_m n$). For simplicity we will assume: all functions φ^γ in (3.14) contain *u* and these *u* may be find from (3.14). Assume the number of independed equations (3.14) are less *r*.

Let us lead to consider *m*-dimentional function $\psi(t,x) = \{\psi^1, \psi^2, \ldots, \psi^m\}$. The components of this function $\psi^j(t,x)$ *j* = 1,2,…,*m* are continious and have the continuous partial derivatives almost everywhere in *T*.

Name this function – charasteric function. Let us lead also the integrable vector-function

$\lambda_1(t), \lambda_2(t), \ldots, \lambda_p(t)$.

Let us take α- function in form

$$\alpha = \int_S \psi^j(\tau,x)\cos(n,t^j)d\tau - \int_P (\psi^j_{t_j} + \psi^j_{x_i}f^i_j + \lambda_\gamma \varphi^\gamma)dt, \tag{3.15}$$

Where *n* is outer normal to surface *S*, *dτ* is element surface *S*. We present the function *J* = *I* + α in form

$$J = A + \int_P B dt, \quad \text{where} \quad A = \int_S \psi^j(\tau,x)\cos(n,t^j)d\tau, \quad B = f_0 - \psi^j_{t_j} + \psi^j_{x_i}f^i_j + \lambda_\gamma \varphi^\gamma. \tag{3.16}$$

Theorem 3.5. Assume $u(t) \in V$. In order to couple *u(t), x(t)* will be the absolute minimum the purpose function (3.12) it is sufficiently* excicting of α-fuction (3.15) such that

1) $\overline{B} = \inf_{x,u \in U} B(t,x,u)$, 2) $\overline{A} = \inf_{x(\tau)} B > -\infty$, 3) $\overline{x}(t), u(t) \in Q$ (3.17)

The proof is identical [2] №7, but in difference from [2] the theorem 3.5 contain the integrability condition.

If $\overline{x}(t), \overline{u}(t) \notin Q$, then $\overline{J}$ is the lower estimation the function (3.12).

If exist the functions ψ, λ and at least one pair $\overline{x}(t), \overline{u}(t)$ satisfying (3.17), then any other pair satisfying (3.17) is minimum of the function (3.12) and any allowable minimum the function (3,12) is satisfying the points 1, 2 (3.17) (consicvently remark 3 §1). The set contains the same or better solution, then $\overline{x}(t), \overline{u}(t)$ is

$$N = \{t, x, u : B(t, x, u) + f_0(t, x, u) \geq \overline{B} + \overline{f}_0\} \quad \text{on} \quad P^* \times U,$$

Assume, functions $f_j^i(t, x, u), \varphi^\gamma(t, x, u)$ are continuous and differentiable. Let us take ψ^j in form $\psi^j = p_{ij}(t) x_i$. Let us denote:

$$H = p_{ij}(t) f_{ij}(t, x, u) - f_0(t, x, u) + \lambda_\gamma \varphi^\gamma(t, x, u).$$

Then p.1 (3.17) of theorem 3.4 we can rewrite: $H(\overline{u}) = \sup_{u \in U} H$ and nessusary condition of minimum (stationarity condition) following from p.2 (3.17) gives:

$$\frac{\partial B}{\partial x_i} = -\frac{\partial p_{ij}}{\partial t_i} - \frac{\partial H}{\partial x_i} = 0, \quad i = 1, 2, \ldots, n. \tag{3.18}$$

§4. Inverse substitution method

A. From previous paragraph we have: if we know the minimum any function on acceptable set, we can get information about solution the Problem 1 and solve one from Problem a, b, c, g the §1.

It is known, that the most direct Problems inf $f_0(x)$ on X^* or

$$\inf \int_{t_1}^{t_2} f_0 dt$$

on Q (i.e. finding the minimum of main Problem) are difficult or do not have the satisfaction solution. However, if purpose function is not in advance definisfed, the solution for this non-definded purpse is finding easy. This is not suprising. In mathematics it has long been known that many inverse problems are solved more easily than direct problems. An example, let us consider the problem of finding the roots of an algebraic equation. n In the general case for $n > 5$ it is solved with difficulty and her decision (roots) not to be expressed in terms radicals. If the roots are given, then the corresponding algebraic equation may be found easy. On the basis of this idea below it is given method to build function for which an admissible element would be the point of absolute minimum on an admissible set: **Inverse Substitution method.** Since we thus have to solve a problem back to the original problem (not find the minimum given function, but find the function for given the minmum or for fiven field). This method is called **the method of reverse lookup.** The method is presented for two cases: problems of the theory of extrema of functions of a finite numbers of variables (p.B) and optimization problems described by ordinary differential equations (p. C).

B. Let us consider usial Problem of minimum the function of finite variables

$$I = f_0(x), \quad f_i(x) = 0, \quad i = 1, 2, \ldots, m < n. \tag{4.1}$$

Let us convert this Problem. Select m componets x and name them main (base). Suppose for definiteness that this is the first components m of the vector x. The rest of components $n - m = r$ denote u_j ($j = 1, 2, \ldots, r$). Again the Problem (4.1) we can re-write

$$I = f_0(x, u), \quad f_i(x, u) = 0, \quad i = 1, 2, \ldots, m < n. \tag{4.2}$$

where x is m - dimentional vector, $x \in X$, $u - r$ - dimentional vector, $u \in U$.

Let us take more simple purpos function $J_1(x, u)$ and find it's the absolute minimum on $X \times U$. This solution may be used for building of sets M, N, P:

$$M = \{x,u : J_1 - f_0 \geq \bar{J}_1 - \bar{f}_0\}, \quad (4.3)$$
$$N = \{x,u : J_1 + f_0 \leq \bar{J}_1 + \bar{f}_0\}, \quad (4.4)$$
$$P = \{x,u : J_1 - f_0 \leq \bar{J}_1 - \bar{f}_0\}. \quad (4.5)$$

Desandvantage this method is next: the some of these sets can not have the admissible elements (i.e. x, u satisfacting $f_i = 0$).

Assume, the limitations $f_i(x,u) = 0$ in (4.2) may be solved abou x:
$$x_i = x_i(u), \quad i = i = 1,2,\ldots,m \quad (4.6)$$

and $x \in X$ for any $\vee u \in U$.

Assume we take simple function $J_1(x,u)$. Substitute in it's the (4.6) and find $\inf_U J_1(x(u),u)$, $\bar{u}$, and (4.6) $\bar{x}$. This solution is analog (4.3)-(4.5). One may be used for finding sets M, N, P. The intersection of these sets with admissible set is not empty. You can take $J_1(x,y,u)$, than $\bar{u} = \bar{u}(y)$. You can use the dependance of M, N, P from y for changing the "size" of these sets. It is clear assesment

$$\Delta = \inf_y \sup_u [J_1(x(u),u) - I(x(u)),u]$$

C. In point 2 §3 we considered the optimization Problem described by conventional differencial equations
$$I = \int_{t_1}^{t_2} f_0(t,x,u)dt, \quad \dot{x} = f_i(t,x,u), \quad i=1,2,\ldots,n, \quad u \in U. \quad (4.7)$$

We was shown: if we take some function $\psi(t,x)$ and find minimum of $\inf_{x,u} B$ in (t_1, t_2) and $\inf_{x_1,x_2} A$, we get the minimum of Problem 1 or the its lower estimation.

Statement of the Problem. Let us to state the Problem 1 the other way: the find the function which matches the function $\psi(t.x)$ and minimum of this function of the admissible set.

Note. Let us note: the offered statement very different from the back problem of variation calculation. In variation calculation, the back problem states next: we have a curve. Find the function, which gives the minimum in this curve. In common case this problem is more difficult then a direct problem.
In our case the minimum curve not given. We find it by given function $\psi(t,x)$.

Theorem 4.1. The minimum function corresponding function $\psi(t,x)$ is

$$J_1 = \int_{t_1}^{t_2} B_1(t,x)dt = \int_{t_1}^{t_2} -\inf_{u \in U}[-\psi_{x_i} f_i(t,x,u) - \psi_t]dt \quad (4.8)$$

And correcponding to it the minimum curve is given by equations

$$\dot{x}_i = f_i[t,x,\bar{u}(t.x,\psi_{x_i},\psi_t)], \quad i = 1,2,\ldots,n, \quad (4.9)$$

where $\bar{u} = \bar{u}(t,x,\psi_{x_i},\psi_t)$ we find from (4.8).

Proof. Write the expression B (see (3.11)) for problem (4.7) and check up condition (3.8) of theorem 3.1:

$$B_2(t) = \inf_{x,u}[B_2(t,x) - \psi_{x_i} f_i(t,x,u) - \psi_t]$$

(4.10)

Obviosly, the (4.10) identically equals zero for $\psi = \psi(t,x)$ from (4.8) and $\bar{x}, \bar{u}$ satisfacting (4.7). If we take as $x(t_2)$ the value $x(t)$, received from (4.9) for t_2, then the point 2 (3.8) disappear and all condition (3.8) of theorem is executed. Theorem is prooved.

Consequence. If $B_1 = f_0(t,x)$, then $x(t)$ getting from (4.10) give the set of the minimal curves for boundary condition $\psi_2 = \psi$. In particular, if the end of curve $x(t)$ from (4.9) match with given boundary conditions, that this curve is minimum curve of Problem 1.

Note. Boundary conditions in the left end can always be performed. For it we must start the integration from the given conditions (4.9). We can perform the boundary condition in the right end the next method. Take in form $\psi(t,x,c)$ where $c - n -$ dimentional constant. Substitute $\psi(t,x,c)$ in (4.9) and select c such that to perform the given end condition in the right end.

Getting numerical function may be used for receiving the set N, P of Theorem 3.3 :

$$N - \{t,x : f_0 + B_1 \le \bar{f}_0 + \bar{B}_1\}, \quad P = \{t,x : B_1 - f_0 \le \bar{B}_1 - \bar{f}_0\},$$
$$\text{where} \quad f_0 = f_0[t,x,\bar{u}(t,x,\psi_x,\psi_t)], \quad \psi(t,x) \text{ is given.}$$

If we find

$$\bar{J} = \psi_2 - \psi_1 + \int_{t_1}^{t_2} \inf_x (f_0 - B_1) dt$$

We get also the lower estimation.

Memo, the assignment $\psi(t,x)$ gives us not single nometical function and its point of minimum. One gives a set of minimums satisfaction the boundary conditions $\psi_2 - \psi_1 = c$.

Note: We can take $\psi(t,x,y)$. Then $B_1(t,x,y)$. If we can select such $\bar{y}(t)$ that $B_1(t,x,\bar{y}) = f_0(t,x)$ and boundary conditions is perfomed, then $\bar{u}(t,x,\bar{y})$ is the optimal synthesis of Problem 1.

D. We also show: how you can find the numerical function for given the syntes of control $u = u(t,x)$.

Equate the given $u = u(t,x)$ to the control findedfrom (4.8). We get the equation in particular derivities

$$u(t,x) = \bar{u}(t,x,\psi_{x_i},\psi_t)$$

(4.11)

Substitute its solution $\psi(t,x)$ and given $u(t,x)$ in (4.8), we find the numerical corresponding function. If $B_1 = f_0(t,x)$ that is synthesis the Problem 1 for the bounded condition $\psi_2 = \psi$.

Possible the other method. We take $u = u(t,x,y)$. Substitute its in (4.8). Then $B_1 = B_1(t,x,c,y)$. We can try using y to reach the identify $f_0 \equiv B_1$ and using c to minimize the nymerical function I.

Example 4.1. Let us consider the task of design the regulator

$$I = \int_{t_1}^{t_2} b_{ij} x_i x_j dt, \quad (4.12)$$

$$\dot{x}_i = a_{ij} x_j + u, \quad 0 \le t \le \infty, \quad (4.13)$$

$$x_i(0) = x_{i,o}, \quad x_i(\infty) = 0, \quad (4.14)$$

where $f_0 = b_{ij}x_ix_j$ is the positive definite form.

Take $u = c_ix_i$, where c_i are constants. Let us to search ψ as the quadratic form $\psi = A_{ij}x_ix_j$ with unknown coefficients. Equate $f_0 \equiv \dot{\psi}$:

$$b_{ij}x_ix_j = A_{ij}x_i(a_{ij}x_j + c_jx_j).$$

Let us equate coefficient in same x_i, x_j in left and right of this equation. We get the set $n(n+1)/2$ the linear inhomogenius equations having the same number of unknown A_{ij}. If the determinant of this system $\Delta \neq 0$, we find A_{ij}. We substitute $f_0 \equiv \dot{\psi}$ in (4.12), integrate and find $I = \psi(\infty,c) - \psi(0,c)$ or using (4.14) $I = -\psi(x_{io},c)$. When we find minimum of this expression for c, we get the optimal syntes. If $-\psi(x,\bar{c})$ is the positive definite form then this function is the Lyapunov function (because $-\dot{\psi} \geq 0$ and the regulator is assimptotic stable.

§5. Method of combining extrema in problems of constrained minimum.

We will show in this paragraph that method combining extrema, considered in §2 the Chapter 1, it is apply in tasks of theory the functions of a finite number of variables (point A) and tasks described the conventional difference equations.

A) Let us again consider the Problem of the theory the functions of a finite number of variables

$$I = f_0(x), \quad f_i(x) = 0, \quad i = 1,2,...,m. \tag{5.1}$$

Write the numerical function

$$J(x,c) = f_0(x) + \beta(x,c) + \alpha_1(x), \tag{5.2}$$

Here $\alpha_1(x)$ is α – function, c is n – dimentional constant.

From condition

$$\inf_{x \in X^*} J(x,c), \tag{5.3}$$

we find $\varphi_1(x^{(1)},c) = 0.$

From condition

$$\Phi(x,c) = \sup_{x \in X^*}[\beta(x,c) + \alpha_2(x)], \tag{5.4}$$

we find $\varphi_2(x^{(2)},c) = 0.$ Solve equations φ_1, φ_2 together with (5.1) (combining equations):

$$\varphi_1(x^{(1)},c) = 0, \quad \varphi_2(x^{(2)},c) = 0, \quad x^{(1)} = x^{(2)}, \tag{5.5}$$

we receive the absolite minimum the Problem 1. The edditive $\beta(x,c)$ selectes so that tasks (5.3), (5.4) are solved easier.

For example, $\alpha_1 = \lambda_i f_i$, $\alpha_2 = \nu_i f_i$. Functions $f_i(x)$, $i = 0,1,...,n$ are continious and difference, the functions $J(x,c)$, $\Phi(x,c)$ have single minimum and maximum for any c. That we have system $(3n + 2m)$ equations with same numbers of unknown magnitudes $\alpha(1)$, $\alpha(2)$, c, λ, ν.

$$J'_{x_j}(x^{(1)},c,\lambda) = 0, \quad \Phi'_{x_j}(x^{(2)},c,\nu) = 0, \quad f_i(x^{(1)}) = 0, \quad f_i(x^{(2)}) = 0, \quad x_j^{(1)} = x_j^{(2)},$$

$$j = 1,2,...,n; \quad i = 1,2,...,m \tag{5.6}$$

We can simplisity this system if we take a vector the dimention (n-m) and use the last equation in (5.6) we exclude $x^{(2)}$. We get the system $(2n+m)$ equations:

$$J'_{x_j}(x^{(1)},c,\lambda) = 0, \quad \Phi'_{x_j}(x^{(2)},c,\nu) = 0, \quad f_i(x^{(1)}) = 0 \tag{5.6'}$$

having $(2n+m)$ unknown valies x, λ, ν, c.

This note is right to systems (5.5), (5.1), which get the form

$$\varphi_1(x,c) = 0, \quad \varphi_2(x,c) = 0, \quad f(x) = 0.$$

Example 5.1 is not include.(see page 52 [2]).

B) Let us to consider the task, described the conventional differential equations:

$$I = \int_{t_1}^{t_2} f_0(t,x,u)dt, \quad \dot{x} = f_i(t,x,u), \quad i = 1,2,...,n, \quad u \in U, \quad x(t_1) = x_1, \quad x(t_2) = x_2 \tag{5.9}$$

Take ψ in form $\psi^{(1)} = p_i^{(1)}(t)\alpha_i^{(1)}$ and create the function

$$B_1 = f_0 + \beta(t,x^{(1)},u^{(1)},z) - p_i^{(1)} f_i^{(1)} - \dot{p}_i^{(1)} x_i^{(1)} = -H^{(1)} - \dot{p}_i^{(1)} x_i^{(1)}.$$

Here $z(t)$ is r – dimentinal function. One can have the limited gaps the first type.

From $\inf_{x,u} B_1$ and (5.9) we find

$$\dot{p}^{(1)} = -H_x^{(1)}, \quad \bar{u}^{(1)} = \bar{u}^{(1)}(t,x^{(1)},p^{(1)},z), \quad \dot{x}^{(1)} = f(t,x^{(1)},u^{(1)}). \tag{5.10}$$

Take $\psi^{(2)} = p_i^{(2)} x_i^{(2)}$ and create the function

$$B_2 = \beta(t,x^{(2)},u^{(2)},z) - p_i^{(2)} f_i^{(2)} - \dot{p}_i^{(2)} x_i^{(2)} = -H^{(2)} - \dot{p}_i^{(2)} x_i^{(2)}.$$

From $\inf_{x,u} B_2$ and (5.9) we find

$$\dot{p}^{(2)} = -H_x^{(2)}, \quad \bar{u}^{(2)} = \bar{u}^{(2)}(t,x^{(2)},p^{(2)},z), \quad \dot{x}^{(2)} = f(t,x^{(2)},u^{(2)}). \tag{5.11}$$

Using the combining equation: $x^{(1)} = x^{(2)}$, $u^{(1)} = u^{(2)}$ we get final:

$$\dot{x} = f(t,x,u^{(1)}), \quad \dot{p}^{(1)} = -H_x^{(1)}, \quad \dot{p}^{(2)} = -H_x^{(2)}, \quad \bar{u}^{(1)}(t,x,p^{(1)},z) = \bar{u}^{(2)}(t,x,p^{(2)},z) \tag{5.12}$$

That is system $3n + r$ equations with $3n + r$ unknown x, $p^{(1)}$, $p^{(2)}$, z. Last equation in (5.12) is the combining equation. The additive function β selecting so that the solution task of finding *inf* and *sup* were simpler.

§6. Generalizing the Theorem 3.1 in case the brocken $\psi(t,x)$.

Theorem 6.1. *Assume there is numerical function $\psi(t,x)$ defined on set $T \times G$, bounded below, piecewise differentiable and piecewise continuous. The function $\psi(t,x)$ and its derivatives can have the breaks the first* types on the limited set $\Phi_s(t_s, x)$, $s = 1,2,...,k-1$ zero measure. This function is such that there is:

1) $\inf_R (F + \psi_k - \psi_0)$, 2) $\inf_{t_s, x \in \Phi_s} (\psi_s^- - \psi_s^+)$, $\bar{t}_s$ f $\bar{t}_{s-1}$, $\bar{t}_k$ f $\bar{t}_{k-1}$, $s = 1,2,...,k-1$,

3) $\inf_{G \times T} B = 0$, 4) $\bar{x}(t), \bar{u}(t) \in Q$.

Then $\bar{x}, \bar{u}$ (are got from points 1 -3) is the absolute minimum the Problem 1.

Here ψ_s^-, ψ_s^+ are value ψ in left and right side (along $\bar{x}(t)$) of the breaks the function ψ and its derivatives.

Proof: From poins 1 – 3 we have

$$\bar{J} = \inf_R (F + \psi_k - \psi_0) + \sum_{s=1}^{k-1} \inf_{t_s, x} (\psi_s^- - \psi_s^+) + \sum_{s=0}^{k-1} \int_{t_s}^{t_{s-1}} \inf_{x,u} B dt$$

On feasible cirves (from Q) the $\bar{J}$ convert in function $I = F + \int_{t_1}^{t_2} f_0 dt$. In this case if we apply the consequence 4 , §1, point 4 of the theorem statement is obviously. Theorem is prooved.

Note. The conditions 3 of Therem 6.1 is sometimes difficult to check up. In this case the requirements 2 - 3 of theorem 6.1 we can change the damage

$$\inf_{t_s} [\inf_x (\psi_s^- - \psi_s^+) + \int_{s-1}^{s} \inf_{G \times U} B dt + \int_s^{s+1} \inf_{G \times U} B dt]$$

One must be checked up in every point t_s, $s = 1,2,...,$k-1.

§7. Optimization the problems described the conventional differential equations having the limitations.

We find minimum A, B in Theorem 3.1, chapter II on the corresponding sets R and $U \times G$. The most widely method of separating the feasible sets is the separation of them from more widely set by equilities and inequilities. In this case, we can solve our problem by the methods the α- and β-functions.

Let us shortly consider the most common cases.

1. Limitations are the equilities

a) Assume the admissible set R is separated by equilities:

$$g_i(x_1, x_2) = 0, \quad i = 1,2,...,l < 2n \tag{7.1}$$

Then the task *inf A* we can change the task

$$\inf_{x_1, x_2} [A + \mu_i(x_1, x_2, z_i) g_i(x_1, x_2)] \tag{7.2}$$

Here μ_i is known functins, z is *l*-dimentional unknown vector. In particular, we can take $\mu_i = z_i$.

b) Let us assume the admissible set $U \times G$ is separated by equilities

$$\varphi_i(t,x,u) = 0, \quad i = 1,2,\ldots,l < r \tag{7.3}$$

Assume, we can find from (7.3) the *l* component the vector *u*. Than the problem $\inf_{G \times U} B$ we can change the problem

$$\inf_{x,u}[B + \lambda_i(t,x,w)\varphi_i(t,x,u)], \tag{7.4}$$

Where λ_i are known function, w_i is *l*-dimantional unknown vector function. In particular, we can take $\lambda_i = w_i$.

c) Assume the admisseble set G is separated by the equilities

$$\varphi_i(t,x) = 0, \quad i = 1,2,\ldots,l < r. \tag{7.5}$$

Differenciate (7.5) full case for *t* and find

$$\varphi_i^{(1)}(t,x,u) \equiv \frac{\partial \varphi_i}{\partial x_j} f_j(t,x,u) + \frac{\partial \varphi_i}{\partial t} = 0, \quad i = 1,2,\ldots,l < n \tag{7.6}$$

If in system (7.6) there is equations do not contain *u*, we differenciate them next time and so on whole we get the the system where all *l* eqution contain u. Assume we can find all *l* components from this system (*l* < *r*).

Than the problem (7.5) is redused to the tasks the point *a, b* in which (7.6) is (7.3), but (7.5) and all equtions (7.6) not contain *u*, are (7.1).

2. Limitations are inequalities (excerpt).

a) Feasible set R is allocated by inequalities:

$$g_i(x_1,x_2) \leq 0, \quad i = 1,2,\ldots,l.$$

Then acording the Teorem 1.4 Chapter 1 we change the problem $\inf_R A$ by problem (7.2) with the additional conditions:

$$\overline{\lambda}_i \overline{g}_i = 0, \quad \overline{\lambda}_i \geq 0 \quad (\text{here} \quad i \quad \text{is} \quad \text{not} \quad \text{sum}) \tag{7.7}$$

b) Feasible set $U \times G$ is allocated by inequalities:

$$\varphi_i(t,x,u) \leq 0, \quad i = 1,2,\ldots,l. \tag{7.8}$$

All inequalities contain u. Then the task $\inf_{U \times G} B$ we change the task (7.4) wuth conditions

$$\overline{\lambda}_i \overline{\varphi}_i = 0, \quad \overline{\lambda}_i \geq 0 \quad (\text{here} \quad i \quad \text{is} \quad \text{not} \quad \text{sum}) \tag{7.9}$$

==

Example 7.1-7.2 are missing (see pages 52-57 [2]).

Part of the text are missing: §8. Optimization of discret system (see pages 58 [2]).; §9. Optimization of systems dependent on intermediate values (see pages 60 [2]).
==

§10. Note on the equivalence of different forms of variational problems

A) In §3 the next problem of minimization was considered

$$I = F(x_1, x_2) + \int_{t_1}^{t_2} f_0(t, x, u) dt, \qquad (10.1)$$

on solution of equations

$$\dot{x}_i = f_i(t, x, u), \quad i = 1, 2, \ldots, n. \qquad (10.2)$$

In the theoretical analysis for the simplicity, we often assume that in (3.1) $F \equiv 0$ or $f_0 \equiv 0$. We show that it does not restrict the generality of our reasoning.

Take

$$I = \int_{t_1}^{t_2} f_0(t, x, u) dt$$

And differentiate it for the variable upper limite t and designate $\dot{x}_{n+1} = f_0$. We get the task

$$I = x_{n+1}(t_1), \quad \dot{x}_i = f_i, \quad \dot{x}_{n+1} = f_0. \qquad (10.3)$$

B) Assume $I = F(x_1, x_2)$. Differenciate it by t and integrate, we get numerial function

$$I = \int_{t_1}^{t_2} (F_{x_i} f_i) dt \qquad (10.4)$$

We can same way to convert (10.1) in (10.4) and in (10.3).

C) Let us to assume the (10.1) and (10.2) depend from constants c_k which must be optimal. Designate $c_k = x_{i+k}$ and add to (3.3) equation $\dot{x}_{n+k} = 0$. We reduced the task having the optimising constants to conventional task.

In practice it is camfortable to solve the problema (10.1), (10.2) with constant parameters. Than to change them (for example the gradient method) so, the function (3.1) decreases.

D) The problem with $f_i(t,x,u)$ which obviously depend from t, we can reduse to problem $f_i(x,u)$ do not depend obviously from t, if to designate $t = x_{n+1}$ and add to (10.1) the equation $\dot{x}_{n+1} = 1$.

Let us to show how the task with the mooving ends t_1 and t_2 we can reduse the task with fix interval of integrate. Take the new variable $t = c\tau$. Than task (10.1), (10.2) having variables t_1 or t_2 was redused in task with fix interval (τ_1, τ_2):

$$I = F + \int_{t_1}^{t_2} c f_0(\tau, x, u) d\tau, \quad x' = c f_i(c\tau, x, u),$$

where the touch means the derivative for τ. The constant $c > 0$ is selected from minimum I.

Application to Chapter II.

1. Theorem 3.1 and known methods of solution the problem described the ordinary differential equations.

From Theorem 3.1 we can to get the conditions which are same with known algorithms of optimal control, for example: Pontriagin principle [3], Bellman equation [4], classical calculus of vatiation [5],

Let us to request additional that function f, ψ have the need continious derivatives.

a) Pontriagin principle. According [2] take $\psi(t,x)$ in form $\psi = p_i(t)\Delta x_i$, where $p_i(t)$ are some differenciable functions t, $\Delta x_i = x_i - \bar{x}_i$. Create the Hamiltonian

$$H = p_i f_i(t,x,u) - f_0(t,x,u).\tag{1}$$

Then $B = -H - p_i x_i$. Necessary condition of the minimum B for x, which follows from p.1 (3.8) of Theorem 3.1 (stationarity condition) is

$$B_{x_i} \equiv -p_i - H_{x_i} = 0, i = 1,2,...,n.\tag{2}$$

Moreover of claim 1 (3.8) we have

$$B(t,x,\bar{u}) = \inf_{u \in U} B(t,x,u) \quad \text{or} \quad \inf_{u \in U}(-H) = -\sup_{u \in U} H \tag{3}$$

Terms and conditions (2), (3) together with (3.3) coincide with the corresponding terms and conditions of the Maximum principle* [1].

b) Belman equation. Assume $x_n \neq 0$. Take all $\lambda_i = 0$ $i = 1,2,...,n-1$ with exeption $\lambda_n = \psi(t,x)/x_n$. Substitute them in (3.9) §3, we get the known Bellman equation [5]

$$\inf_{u \in U}(f_0 - \psi_{x_i} f_i - \psi_t) = 0 \tag{4}$$

Boundary condition for them is $A = const$. Solution of this equation is the field of all optimal trajectories.

c) Classical calculus of variation. From claims 1, 2 Theorem 3.1 easy to get the conditions of a relative minimum coinciding with the relevant terms of the calculus of variations [6].

Let us assune U is the open area, $\dot{x}(t), u(t)$ are continiosly, $f_i(t,xu)$ have continious partial deriveties up the third order. Take $\psi = p_i(t)\Delta x_i$. From (3) that at minimum

$$B_{u_i}(t,x,u) = -H_{u_i}(t,x,u) = 0, \quad i = 1,2,...,r,\tag{5}$$

Equtions (2),(4) equal the conventional Eiler-Lagrange equations [5] §2 p.1. From [3] also follow

$$-H_{u_i u_j} \delta u_i \delta u_j \geq 0, \quad i,j = 1,2,...,r.\tag{6}$$

That matches with Klebs condition - condition of weak relative minimum.

From (3), one can obtain a condition that coincides with the Weystrass condition.
В самом деле, если B выбрано согласно (3), то

$$B(t,x,u) - B(t,x,\overline{u}) \geq 0, \quad i = 1,2,...,r, \qquad (7)$$

Let us to take Ψ in form $\Psi = p_i(i)\Delta x_i$. Take in our attention the (1), the inequility (7) we can rewrite in form

$$f_0(t,x,u) - p_i(t)f_i(t,x,u) - f_0(t,x,\overline{u}) - p_i(t)f_i(t,x,\overline{u}) \geq 0, \quad i = 1,2,...,r, \qquad (8)$$

Here u is any, the $\overline{u}$ is values corresponding to $\inf_{u \in U} B$.

Let us to add in (8) the values Identical zeros

$$p_i[\dot{x}_i - f_i(t,x,u)] = 0, \quad p_i[\dot{x}_i - f_i(t,x,\overline{u})] = 0. \qquad (9)$$

Then

$$f_0 - p_i(\dot{x}_i - f_i) - \overline{f}_0 - p_i(\dot{x}_i - \overline{f})_i - p_i f_i + p_i \dot{x}_i \geq 0, \qquad (10)$$

Here $\overline{f}_i = f_i(t,x,\overline{u})$. Let us to write known in variable calculation the Lagrange function.

$$\overline{F} = f_0(t,x,\overline{u}) + p_i[\dot{x}_i - f_i(t,x,\overline{u})], \qquad (11)$$

where the role of the uncertain Lagrange multipliers plays $p_i(t)$. According (9) $p_i = \partial F / \partial \dot{x}_i$,
Using (10) and (11), we get

$$F - \overline{F} - (\dot{x}_i - \dot{\overline{x}}_i)\overline{F}_{\dot{x}_i} \geq 0 \qquad (12)$$

where is $F = f_0 + p_i(\dot{x}_i - \dot{\overline{x}}_i)$. Inequality (12) is same with Wirstrasse condition of strong relative minimum.

From p.2 (3.8) we can get condition same the transversality condition of variation calculation. Assume the set R is the space E_n. Then the stationary condition following from p.2 (3.8) of theore 3.1 gives condition same the transversality condition:

$$\left[\frac{\partial F}{\partial x_i} + \frac{\partial \psi}{\partial x_i}\right]_t = 0, \quad i = 1,2,...,n, \qquad (13)$$

Condition same the Yakobi condition of relative minimum we can get from p.1, 2 (3.8) of Theorem 3.1. For simplicity, we consider the case of fixed ends. Calculate d^2J we get

$$d^2J = d\int_{t_1}^{t_2} B dt = \int_{t_1}^{t_2} (B_{x_i x_j}\delta x_i \delta x_j + 2B_{x_i u_\beta}\delta x_i \delta u_\beta + B_{u_\beta u_\gamma}\delta u_\beta \delta u_\gamma) dt \geq 0 \qquad (14)$$

$$i,j = 1,2,...,n, \quad \beta,\gamma = 1,2,...,m,$$

Where $\delta x_i, \delta u_\beta$ connected the variation equations

$$\delta\alpha_i = f^i_{x_j}\delta x_j + f^i_{u_\beta}\delta u_\beta, \quad i,j = 1,2,...,n, \quad f_i = f^i.$$

You can see, the equation under integral in (14) is same with the second variation from F (3.7), if $\Psi = p_i(i)\Delta x_i$.

2. Getting from α – function the method "fine".

Let us to consider the problem of search the minimum the varable functions

$$J = f_0(x), \quad f_i(x) = 0, \quad i = 1,2,...m \text{ р } n. \qquad (1)$$

Take α – functionin form

$$\alpha = a_i f_i^2, \qquad (2)$$

where a_i – constants, $a_i \geq 0$. Evidently, that (2) is α – function, because on accepteble x it equals 0. Build the generalized function

$$J = f_0 + a_i f_i^2, \qquad (3)$$

It is known, for certain conditions for $a_i \to \infty$ the minimum of the general function strives for minimum of function (1).

But from theorem 1.4 we get a new information: the minimum of general function (3) for any α is low estimation function (1).

B) Consider the optimization problem described by conventional differencial equations

$$I = F(x_1, x_2) + \int_{t_1}^{t_2} f_0(t,x,u)dt, \quad x'_i = f_i(t,x,u), \quad i=1,2,...,n \qquad (4)$$

and detail described in §3 p.B.

Take α – function in form

$$\alpha = \int_{t_1}^{t_2} \frac{a_i}{2} [\dot{x}_i - f_i(t,x,u)]^2 dt, \quad i=1,2,...,n \qquad (5)$$

where $a_i > 0$. Search the minimum of function

$$I = F + \int_{t_1}^{t_2} \{f_0 + \frac{a_i}{2}[\dot{x}_i - f_i(t,x,u)]^2\}dt, \quad i=1,2,...,n \qquad (6)$$

For solution this problem we apply the theorem 3.1. Introduce designations
$\dot{x}_i = v_i$, $i=1,2,...,n$.

Here v_i new control. Then general function is

$$I = F + \psi|_{t_1}^{t_2} + \int_{t_1}^{t_2} [f_0 + \frac{a_i}{2}(v_i - f_i)^2 - \psi_{x_i} v_i - \psi_t]dt = A + \int_{t_1}^{t_2} B dt, \quad i=1,2,...,n \qquad (7)$$

Assume the ends of x_i are fixed. Let us take $\psi = p_i(t)x_i$. Substitute it in (7). From condition $\inf_{x,u,v} B$ we get (U is open):

$$B_{v_i} = a_i(v_i - f_i) - p_i = 0 \quad \text{or} \quad v_i = f_i + \frac{B}{a_i}, \quad \dot{x}_i = f_i + \frac{B}{a_i}, \qquad (8)\text{-}(9)$$

Using (8) we find

$$B_{x_j} = \frac{\partial f_0}{\partial x_j} + a_i(v_i - f_i)\left(-\frac{\partial f_i}{\partial x_j}\right) - p_i = \frac{\partial f_0}{\partial x_j} - p_i \frac{\partial f_i}{\partial x_j} - \dot{p}_i = 0,$$

$$B_{u_k} = \frac{\partial f_0}{\partial u_k} - p_i \frac{\partial f_i}{\partial u_k} = 0. \qquad (10)\text{-}(11)$$

Using the **Hamiltonian** definition $H = p_i f_i - f_0$ we get

$$\dot{p}_i = -H_{x_i}, \quad i=1,2,...,n, \quad H_{u_k} = 0, \quad k=1,2,...,k \qquad (12)$$

We see, that (12) for function (6) is same the connection system the maximum principal and the right parts of the connetion equations (4) is different only the eddition p_i/a_i (see (9)).

From here we can see (see (9)) that in case of the limited $p_i(t)$ on (t_1, t_2) for $a_i \to \infty$, the point of minimum (6) strives for the minmum the point of function (4).

From theorem 1.4 we have also the new result: minimum of function (6) for any $a < \infty$ is lower estimation of function (4).

3. Design of the function ψ by solution the integro-differential equation

Let us take take $\psi(t,x)$ in form

$$\psi(t,x) = \int_{\bar{x}_i}^{x_i} \psi_{x_i}(t,x)dx_i \qquad (1)$$

where $\psi_{x_i} = \partial\psi/\partial x_i$. Here all omponents of vector x, exept x_i in integration are parameters. Assume functions ψ_{xi} are continious and ψ_{xii} are exist. Then

$$\psi_t(t,x) = \int_{\bar{x}_i}^{x_i} \psi_{x_i t}(t,x)dx_i \quad (\psi_t = \frac{\partial\psi}{\partial t}, \; \psi_{x_i t} = \frac{\partial^2 \psi}{\partial x_i \partial t}) \qquad (2)$$

Substitute ψ_t in (4) Application 1 we get integro-differential equation

$$\inf_U (f_0 - \psi_{x_i} f_i - \int_{\bar{x}_i}^{x_i} \psi_{x_i}(t,x)dx) = 0$$

If we solve it we find all fild the optimal trajectoris.

4. Common Principle of Reciprocity in variational problems, discribed by the ordinary differential equations

A) Assume in (3.4) $f_0 \equiv 0$ *) and $F = F(t_1, x_1, x_2)$. Assume, we solve the partial differential equation

$$\inf_u [-\psi_{x_i} f_i(t,x,u) - \psi_t] \qquad (1)$$

with end condition $\psi(t_2, x_2) = 0$. Then the general function is

$$J = F(t, x_1, x_2) - \psi(t_1, x_1) . \qquad (2)$$

---
*) This does not violate the generality of the variational problem.

The conditions of therem 3.1 are a single condition

$$\inf_u J(t_1, x_1, x_2) \qquad (3)$$

If we take different F and R *) and solve (3) for any from these function, we can find the value of absolute minimum for any given end conditions.

---
*) Set R can have in given case the variable t_1.

Thus, the solution of equation (1) reduces variational problem for any function to problem of conditional minimum of a function of a finite number of variables.

B) Assume in (3.4) $f_0 \equiv 0$, and $x(t_i)$ are given. Let us be given a function $\psi(t,x,y)$, values $y(t1)$. Using $\inf_{x,u} B$ on (t_1, t_2) and equation (3.3) we find $x(t_2), y(t_2)$. Assume $\bar{x}(t)$ are allowable. Then general function is

$$J = F(x_1, x_2) + \psi(t_2, x_2, y_2) - \psi(t_1, x_1, y_1) + c, \quad y_1 = y(t_1), \quad y_2 = y(t_2).$$

In this case the conditions of theorem 3.1 is single condition

$$\inf_R J(x_1, x_2) . \qquad (4)$$

Let us take different F. If values $\bar{x}_1, \bar{x}_2$ from (4) are same values $x(t_1), x(t_2)$ and (3.3), then it is absolute minimum for gven function. If no, then with according the General principal of repciprocity the magnitude (4) gives the lower estimation for given function or end conditions.

Assume we take $\psi(t,x)$, find $\bar{x}(t), \bar{u}(t)$ from $\inf_{x,u} B$ and it turned out that they did not satisfy the equetions (3.3). Then (4) gives only the lower estimation for any function.

Let us to consider the important case when $\psi = \sum_{i=1}^{n} x_i, y_i$. In this case

$$J = F(x_{i1}, x_{i2}) + \sum_{i=1}^{n} x_{i2}, y_{i2} - \sum_{i=1}^{n} x_{i1}, y_{i1} + c. \quad (5)$$

Here we use the notations $x_i(t_1) = x_{i1}$, $y_i(t_1) = y_{i1}$ and so on.

Assume we want to find the minimum of coordinate $x_1(t_2)$, i.e. $F = x_{1,2}$ for condition, that all rest coordinate are diven. Solve boundary-value problem, i.e. select such quantities y_{i1} so that the values of x_{i2} coincide with the given and $y_{i2} = -1$. Other words if we solved the given optimal problem. In this case this solution is the optimal solution for any function having x_{i2}, y_{i2}, x_{i1}, y_{i1} (no sum for j) for condition: all rest have the end values x_{j1}, x_{j2} ($j \ne i$). In fact, the values of $y_{i1,2}$ are determined up to a constant factor a (the system $\dot{\&} = -H_{x_i}$ is homogeneous). Assume $F = -x_{i2}y_{i2}$ or $F = x_{i1}y_{i1}$ we get that condition $\inf J$ or $x_{i2}y_{i2}$ or $-x_{i2}y_{i2}$ carry out because J not depend from these values. In this way when in (5) $y_{i2} < 0$ then correcponding coordinate x_{i2} reaches the minimum. But if $y_{i2} > 0$ then it is maximum. For x_{i2} it is conversely. This solution will be minimum also for function

$$F = \frac{1}{2} y_{i2} x_{i2}^2, \quad y_{i2} > 0$$

If end value are different from given, the equation (5) gives the lower estimation.

5. Application the theory α – function to the problem of the relative conditional minimum in the theory of functions of a finite number of variables

Assume we need to find the minimum of problem

$$I = f_0(x), \quad f_i(x) = 0, \quad I = 1,2,\ldots,m. \quad (1)$$

Here $x - n -$ dimentinal vector ($m < n$), $f(x)$ Is continuous and twice differentiable.
Let us to apply the theorem 2.5. We will find $\alpha -$ function in form

$$\alpha = (a_i + \frac{1}{2} b_{ij} \Delta x_j) f_i(x), \quad a,b = const, \, j = 1,\ldots,n, \quad \Delta x_j = x_j - \bar{x}_j \quad (2)$$

Write $J = I + \alpha$ and calculate the first differencial, we get

$$dJ = \left[\frac{\partial f_0}{\partial x_\alpha} + (a_i + \frac{1}{2} b_{ij} \Delta x_j) \frac{\Im f_i}{\partial x_\alpha} \right] \delta x_\alpha + \frac{1}{2} b_{ij} f_i(x) \delta x_j, \quad \alpha, j = 1,\ldots,n.$$

From $dJ = 0$ and arbitrariness δx_j in point of relative minimum $\bar{x}_1 \subset X^s$, we get the system of equations

$$\frac{\partial f_0}{\partial x_\alpha} + a_i \frac{\partial f_i}{\partial x_\alpha} = 0, \quad i = 1,\ldots,m, \quad \alpha = 1,\ldots,n. \quad (3)$$

From this system and (1) we find $\bar{x}_j, \bar{a}_i$. Calculate the second differencial in point $\bar{\alpha}$:

$$d^2 J = \left[\frac{\partial^2 f_0}{\partial x_\alpha \partial x_k} + a_i \frac{\partial^2 f_i}{\partial x_\alpha \partial x_k} + b_{ik} + \frac{\partial f_i}{\partial x_\alpha} \right]_{\bar{x}} \delta x_k \delta x_\alpha = c_{k\alpha} \delta x_k \delta x_\alpha \quad (4)$$

Note: the coefficients of quadratic form (4) are different from coefficients of usial quadratic form (example, in Lagrange method) by additive $b_i = \dfrac{\partial f_i}{\partial x_k}$ having $n \times m$ arbitrary constants b_{ik}. If we can choose them so that the form (4) becomes positive definite, then $\bar{x}$ is point of local minimum. For this purpose one can, for example, find at least one solution $b_{i\alpha}$ of the system of linear inequalities that follow from the Sylvester criterion.

Example. Find the minimum $I = x + y$ for condition $x^2 + y^2 = 2$.
Let us to create system (3):
$$1 + 2x\, a_1 = 0,\ \ 1 + 2y a_2 = 0,\ \ x^2 + y^2 = 2.$$
From this we find the points of extremum
$$\bar{x} = -1,\ \bar{y} = 1,\ a_1 = 1/2 \quad \text{and} \quad \bar{x} = 1,\ \bar{y} = 1,\ a_1 = -1/2.$$
Calculate coefficients $c_{k\alpha}$ in (4): $c_{11} = 2a_2 + xb_{21}$, $c_{21} = 2yb_{22}$, $c_{12} = 2xb_{12}$, $c_{22} = 2a_1 + 2yb_{12}$. From Sylvester criterion we have $c_{11} > 0$, $c_{11}c_{22} - c_{12}c_{21} > 0$. In the first point we have
$$2a_1 - 2b_{22} > 0,\ \ (2a_1 - 2b_{11})(2a_1 - 2b_{22}) - 4b_{21}b_{12} > 0.$$
One possible solution of these inequalities is: $b_{22} = 0$, $b_{12} = 1/3$. Consequently the point (-1-1) is mnimum. Similarly it is possible to show the point (1, 1) is point the maximum.

Exercises for α – function

Using the method α – function, find quasi-minimum has accuracy less 5%.

<u>Decignation:</u> If $\psi(\bar{x}) \neq 0$, we select x_i, few different from $\bar{x}$, but allowable $\psi(x_i) = 0$ and compare $I(x_i)$ to the lower estimation $J(\bar{x})$.

1. $I = 2y^2 - 2x - 2x\cos xy,\ \ \varphi = x + \dfrac{1}{x} - \cos xy = 0.$
   Answer: $\bar{J} = 0,\ \bar{x} = 1,\ \bar{y} = 0,\ \bar{\varphi} = 0.$

2. $I = x^2 - 2x + yx^7 - y^9 - y,\ \ \varphi = x^7 - y^8 - x - y = 0.$
   Answer: $\bar{J} = -1,\ \bar{x} = 1,\ \bar{y} = 0,\ \bar{\varphi} = 0.$

3. $I = \dfrac{8}{x} + \dfrac{x}{y} + y|x-4| - x|y-2|,\ \ (x > 0, y > 0),\ \ \varphi = -\dfrac{1}{x} + \dfrac{|x-4|}{x} - \dfrac{|y-2|}{y} = 0.$
   Answer: $\bar{J} = 6,\ \bar{x} = 4,\ \bar{y} = 2,\ \bar{\varphi} \neq 0,\ \bar{I}(3,2) = 6\dfrac{1}{7} \geq 6.$

4. $I = \dfrac{y-x}{\sqrt{1+x^2+y^2}} + \dfrac{|\sin\tfrac{\pi}{2}xy|}{\sqrt{3}},\ \ \varphi = \dfrac{\sqrt{1+x^2+y^2}}{\sqrt{3}}|\sin\tfrac{\pi}{2}xy| - 1 = 0.$
   Answer: $\bar{J} = -\sqrt{3},\ \bar{x} = 1,\ \bar{y} = -1,\ \bar{\varphi} = 0.$

5. $I = x^2 + y^2 - 2x + x^3 + y^3 x + x^4 y - 9,\ \ \varphi = x^2 + y^3 + z^4 + y - 1 = 0.$
   Answer: $\bar{J} = -10\dfrac{1}{3},\ \bar{x} = \dfrac{2}{3},\ \bar{y} = -\dfrac{1}{3},\ z = 1,\ \varphi \neq 0,\ \bar{I}(0,0,1) = -10.$

6. $I = \dfrac{z^2}{y} - \dfrac{5}{4}x\cos\pi xyz + \dfrac{2}{z} + 6\ \ (x > 0,\ y > 0,\ z > 0),\ \ \varphi \equiv 5\cos\pi xyz + \dfrac{y^2}{x^2} + 4 = 0,$
   Answer: $\bar{J} = 10,\ \bar{x} = \dfrac{1}{2},\ \bar{y} = -1,\ z = 1,\ \varphi \neq 0,\ \bar{I}(1,1,1) = 10\dfrac{1}{2} > 10.$

Example of solution

$I = x^2 - x + y^2 - y \sin 0.5\pi xy + 20, \quad \phi = \sin 0.5\pi xy - 1 = 0.$

Select $\alpha = \lambda(x,y)\phi(x,y)$ so that we can easy find minimum $J(x,y)$. Take $\alpha = y\phi$. Then

$$J = x^2 - x + y^2 - y + 20, \quad \text{Solution} \quad \bar{x} = 1/2, \quad \bar{y} = 1/2, \quad \bar{J} = 19\frac{1}{2}.$$

That is, our decision is not permissible. But it is estimation from bottom. If the function changes smoothly, you can try to find a valid value close to the found point and compare it with the estimate.

Take $x = 1$, $y = 1$. Function $\phi(1,1) = 0$. Consequently this solution $x = 1$, $y = 1$ is permissible. The function $I(1,1) = 20 > 19.5$. From the inequality $0.5/19.5 < 0.05$ it follows that $x = 1$, $y = 1$ can be taken as a quasi-optimal solution.

References to Chapter II.

1. Bolonkin A.A., Post-doctor thesis "New Methods of Optimizations and their Applications in Control Systems". Leningrad Polytechnical Institute, 1969, 285 ps. (In Russiam: Болонкина А.А.: Новые метоы оптимизации и их применение в задачах динамики управляемых систем. **Докторская диссертация**, ЛПИ 1969г).
   https://drive.google.com/file/d/0BzlCj79-4Dz9YTJOUHVhR1FZUVE/view?usp=drive_web Dissertation Optimization 1-2 9 30 15.doc, http://viXra.org/abs/1511.0214 , http://viXra.org/abs/1509.0267 , Part 1; http://vixra.org/abs/1509.0265 Part2,
   https://archive.org/details/NewMethodsOfOptimizationAndItsApplication.Part1inRussian, https://www.academia.edu/s/2a5a6f9321?source=link, http://www.twirpx.com,

2. Bolonkin A.A., New Methods of Optimization and their Application, Moscow, MVTU, 1972, 220 ps. (Russian). Болонкин А.А., **Новые методы оптимизации и их применение.** МВТУ им. Баумана, 1972г., 220 стр. (См. РГБ, Российская Государственная Библиотека, Ф-861-83/1809-6**).**
   http://vixra.org/abs/1504.0011 **v4.**   http://viXra.org/abs/1501.0228, (v1, old) , http://viXra.org/abs/1502.0137 **v3;** http://viXra.org/abs/1502.0055 v2; https://archive.org/details/BookOptimization3InRussianInWord20032415 v2, https://archive.org/details/BookOptimization3InRussianInWord20032415_201502 v3, https://archive.org/details/BookOptimizationInRussian **(old),** http://www.twirpx.com/file/1592607/ v2. http://www.twirpx.com/file/1605604/?mode=submit v3, https://www.academia.edu/11054777/ v.4.

3. Potriagin V.G., etc., Mathematical thory of optimal processes, Moscow, Fizmatgis, 1961 (in Russian).

4. Bellman R., Dinamic Programming, Mosow, Public house of Foring Literature, 1960 (inRussian).

5. Bliss G.A., Lectures in variable calculatuin, Public house of Foring Literature, 1960 (inRussian).

6. Picone Mauro, Criteri sufficieenti

7. Болонкин А.А., **Новые методы оптимизации и их примеение в задачах динамики управляемых систем.** Автореферат диссертации на соискание ученой степени доктора технических наук. Москва, ЛПИ, 1971г., 28 стр. http://viXra.org/abs/1503.0081, 3 11 15.   http://www.twirpx.com , https://archive.org ?(не загружается?>7 Mb?) http://samlib.ru/editors/b/bolonkin_a_a/ , http://intellectualarchive.com/ #1488. https://independent.academia.edu/AlexanderBolonkin/Papers,

8. Болонкин А. А., **Об одном методе решения оптимальных задач.** Известия СО Академии наук СССР, вып.2, № 8, июнь 1970 г. http://www.twirpx.com/file/1837179/ , http://viXra.org/abs/1512.0357 , http://vixra.org/pdf/1512.0357v1.pdf ,   https://www.academia.edu , https://archive.org/download/ArticleMethodSolutionOfOptimalProblemsByBolonkin

9. **List #1 Bolonkin's publications in 1965-1972**.(in Russian).
   https://archive.org/details/No1119651972
10. Bolonkin A.A. ,"Principle of Extention and Yakoby codition of variable calculation", Report of Ukraine Science Academy #7, 1964 (ДАН УССР, №7, 1964)(in Ukraine).

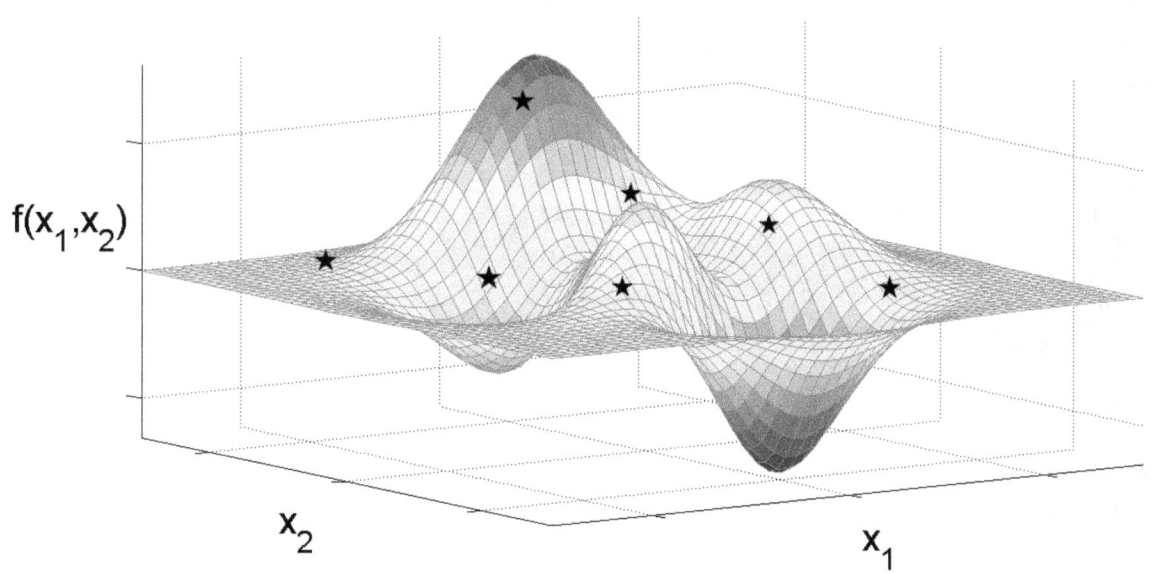

Attachments:

Chapter 12.
Optimal Thrust Angle of Aircraft.

Summary.

The optimal angle for an aircraft's thrust vector is derived from first principles. Two equations are shown to encompass six different flight regimes. The main result for take-off and landing is that the optimal thrust angle in radians approximately equals the coefficient of rolling friction. For climb, cruise, turn and descent, the optimal thrust angle equals the arctangent of the ratio of drag coefficient to lift coefficient. The second result differs from the well-known result that optimal thrust angle equals the arctangent of the partial derivative of drag with respect to lift. The author discusses this difference.

Nomenclature

B	= artificial function, $\lambda dx/dt - H$
C_D	= drag coefficient
C_L	= lift coefficient
D	= drag
d	= take-off or landing distance
E	= aircraft efficiency, C_L/C_D
F	= fuel consumption
f	= performance function
g	= n-dimensional vector constraint function
g_0	= acceleration due to gravity
H	= Hamiltonian, $-f + \lambda g$
h	= altitude
I	= performance index
L	= lift force
M	= aircraft mass
OTA	= Optimal Thrust Angle
q	= dynamic pressure, $\rho V^2 / 2$
R	= range
S	= wing area
t	= time
T	= thrust or time of flight
T_f	= friction force
u	= m-dimensional control vector
V	= aircraft speed
W	= aircraft weight
w_f	= specific fuel consumption
x	= n-dimensional state vector
λ	= n-dimensional Lagrange multiplier

γ = thrust angle
μ = friction coefficient
ψ = specific function
ϕ = roll angle

Introduction

Aircraft designers must determine the angle of the thrust vector relative to the main horizontal flight direction. When this angle is positive (up from the horizontal plane), an additional lift force is generated, but at the expense of horizontal thrust. In this paper, the optimal thrust angle is derived, using both classical methods and an alternative optimization method developed by the first author.[1,2] Many methods of deflecting the nozzle exhaust stream of rocket engines to provide thrust vector control have been investigated, including jet vanes, gimbaled or swiveled nozzles, and extendable nozzle deflectors.[3,4,5,6] Jet vanes have been widely applied for the control of solid rocket engines and for early liquid-rocket engines, including the German V-2 missile.[7] Reference 8 presents metrics for assessing the performance of fighter aircraft implementing thrust vector control.

References 3 and 9 are most closely related to this paper. In Reference 3, the authors use numerical calculations to search for the optimal thrust angle, whereas in this paper the focus is theoretical, rather than numerical. In Reference 9, Miele presents a basic theory for analyzing the optimum flight paths of rocket-powered vehicles. Miele simultaneously optimizes the time history of lift, thrust modulus and thrust direction, and states that the optimal thrust angle equals the arctangent of the partial derivative of drag with respect to lift. In this paper, we provide theory and formulas for the OTA for six primary flight regimes of any aircraft type. The formulas provided are accurate for stable flight conditions, but may be sub-optimal during high dynamic maneuvers. The six flight regimes are listed below, each with one or more optimization objectives.

1. Take-off, to minimize take-off ground run.
2. Climb, to minimize fuel consumption.
3. Cruise, to minimize fuel consumption or to maximize range.
4. Turn, to minimize fuel consumption or to minimize turn time.
5. Descent, to minimize fuel consumption or to maximize range.
6. Landing, to minimize landing roll.

General Methodology

Consider the problem of minimizing a performance index *I*, where

$$I = \int_0^T f(t,x,u)dt \tag{1}$$

We wish to minimize *I* with respect to *x* and *u*, subject to the dynamic constraint

$$\dot{x} = g(t,x,u) \tag{2}$$

We assume an initial condition, $x(0)$, is known. Following the approach described in Chapters 1, §4B, 2 or in Reference 1, we define an artificial α - function in form

$$B = f - \frac{\partial \psi}{\partial x} g(t,x,u) - \frac{\partial \psi}{\partial t} \tag{3}$$

In particular, the function ψ may be defined by

$$\psi = \lambda(t) \cdot x \qquad (4)$$

If we find

$$\min_{x,u} B = \min_{x,u}[f(t,x,u) - \lambda(t)g(t,x,u) - \dot{\lambda}(t)x] \qquad (5)$$

then the values of x and u that minimize B, subject to the constraint given in Eq. (2), are optimal control and state vectors for the problem stated in Eq. (1).

For readers are not friendly with Method of Deformation, we can also solve this problem by a conventional method[10] using the Hamiltonian for this problem (as it is shown in Chapter 2 most conventional methods may be received from Method of Deformation), which is given by

$$H(t,x,u,\lambda) = -f(t,x,u) + \lambda(t) \cdot g(t,x,u), \qquad (6)$$

and $\dot{\lambda} = -\dfrac{\partial H}{\partial x}$

As it is shown in Chapters 1 – 2, Hamiltonian is part of particular function B, which is particular case of α – function and α – function is particular case of β – function

$$\min_{x,u} B = \min_{x,u}[f(t,x,u) - \lambda(t)g(t,x,u) - \dot{\lambda}(t)x] = \min_{x,u}[-H - \dot{\lambda}(t)x] .$$

In B we find the minimum for x, u. In conventional method we find the maximum of H only for u.

The values of u which maximize the Hamiltonian, subject to the constraint in (2), are optimal control vectors for (1). That is

$$\overline{H} = \max_{u} H(t,x,u,\lambda) . \qquad (7)$$

When the process does not change with time, we have a more straightforward problem:

Minimize a performance index I, defined by

$$I = \int_0^T f(x,u)dt \qquad (8)$$

with respect to x and u, subject to the dynamic constraint

$$\dot{x} = g_i(x,u), \quad \text{for } i = 1,2, ..., n \qquad (9)$$

$$H(t,x,u,\lambda) = -f(x,u) + \lambda(t)g(x,u) \qquad (10)$$

$$\overline{H} = \max_{u} H(t,x,u,\lambda), \qquad \overline{u} = u(t,x) \qquad (11)$$

The parameter λ is an n-dimensional unknown Lagrange multiplier and $\overline{u}$ is the optimal control. Eqs. (4) through (6) give the system of equations

$$\frac{\partial B}{\partial u_j} = \frac{\partial H}{\partial u_j} = 0, \quad j = 1,2,...,m; \quad \frac{\partial B}{\partial x_i} = \dot{\lambda}_i(t) + \frac{\partial H}{\partial x_i} = 0, \quad i = 1,2,...,n \tag{12}$$

These equations are equivalent to conventional principle of maximum[10];

$$\dot{\lambda}_i(t) = -\frac{\partial H}{\partial x_i}, \quad i = 1,2,...,n; \quad \frac{\partial H}{\partial u_j} = 0, \quad j = 1,2,...,m. \tag{13}$$

These equations, together with Eq. (2), allow us to find an extreme of the Hamiltonian H, which is optimal if the appropriate second order sufficient conditions for optimality are satisfied.

Optimal Thrust Angle for Take-off and Landing

For take-off, the performance index is the take-off distance, described by

$$d = \int_0^T V dt. \tag{14}$$

The aircraft speed serves as the performance function. The dynamic constraint on acceleration is given by

$$\dot{V} = \frac{1}{M}(T\cos\gamma - D - T_f) \tag{15}$$

as illustrated in Fig. 1.

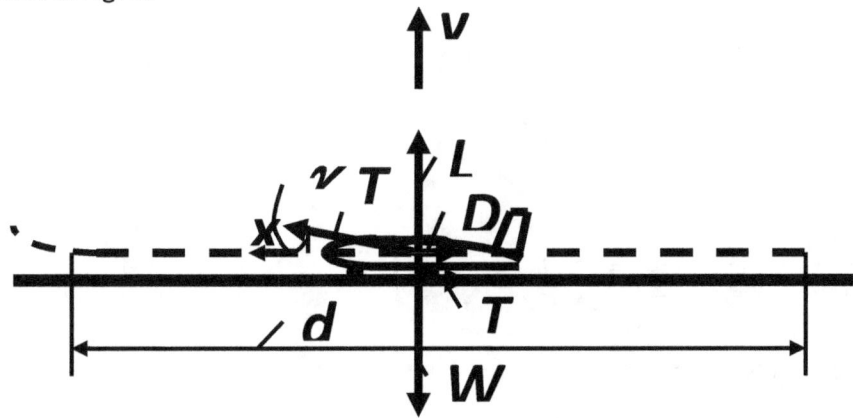

Fig. 1 Take-off

The friction force is given by

$$T_f = \mu \cdot (W \cdot g_0 - L - T\sin\gamma) \tag{16}$$

We know from aerodynamics and trigonometry that

$$L = C_L qS, \quad D = C_D qS, \quad \sin\gamma = \sqrt{1 - \cos^2\gamma} \tag{17}$$

We consider only the positive root, but the result is the same for the negative root.
To simplify subsequent calculations, make the substitution

$$u = \cos\gamma \qquad (18)$$

so that

$$\sin\gamma = \sqrt{1-u^2} \qquad (19)$$

Substituting Eqs. (16) through (19) into (15) yields

$$\dot{V} = \frac{1}{M}[T \cdot u - D - \mu \cdot (W \cdot g_0 - L - T \cdot \sqrt{1-u^2})] \qquad (20)$$

which leads to function B or the Hamiltonian H

$$B = f + \dot{\lambda}(t) \cdot V + \lambda(t) \cdot \dot{V} = V + \dot{\lambda}(t) \cdot V + \lambda(t) \cdot \frac{1}{M}[T \cdot u - D - \mu \cdot (W \cdot g_0 - L - T\sqrt{1-u^2})]$$

$$= \dot{\lambda}(t) \cdot V + H, \qquad (21)$$

where

$$H = V + \lambda(t)\frac{1}{M}[T \cdot u - D - \mu \cdot (W \cdot g_0 - L - T\sqrt{1-u^2})] \qquad (22)$$

To find the minimum of B over all admissible u, the necessary condition is that the partial derivative is equal to zero, that is,

$$\frac{\partial B}{\partial u} = 0 \qquad (23)$$

or

$$\frac{\partial B}{\partial u} = \frac{\lambda(t) \cdot T}{M}\left[1 - \frac{\mu \cdot u}{\sqrt{1-u^2}}\right] = 0 \qquad (24)$$

If M, T, and $\lambda \neq 0$, then from (24), it must be true that

$$\mu \cdot u = \sqrt{1-u^2} \qquad (25)$$

or

$$\mu^2 \cdot u^2 = 1 - u^2 \qquad (26)$$

so that the final result from (26) is

$$u = \pm\frac{1}{\sqrt{1+\mu^2}} \qquad (27)$$

Returning to the original notation, we have the thrust angle as a function of the coefficient of friction,

$$\cos \gamma = \pm \frac{1}{\sqrt{1+\mu^2}} \quad \text{or} \quad \gamma = \cos^{-1}\left(\frac{\pm 1}{\sqrt{1+\mu^2}}\right) \text{ radians} . \qquad (28)$$

We can use the trigonometric identity

$$\cos \gamma = \frac{1}{\sqrt{1+\tan^2 \gamma}} \qquad (29)$$

to get our final result,

$$\boxed{\tan \gamma = \pm \mu} \qquad (30)$$

or, for small µ, say µ < 0.2, we have the design rule-of-thumb that

$$\boxed{\gamma \cong \pm \mu} \qquad (31)$$

where γ is in radians.

The sign of γ depends on our goal, minimization or maximization of the function, as well as the sign of λ and T in Eq. (24). Clearly, the thrust must have a forward direction for aircraft take-off, and the angle γ must be positive. Similarly, for landing, the thrust must have a backward direction to brake the airplane, and the angle γ must be negative, pushing the airplane to the ground, as illustrated in Fig. 2.

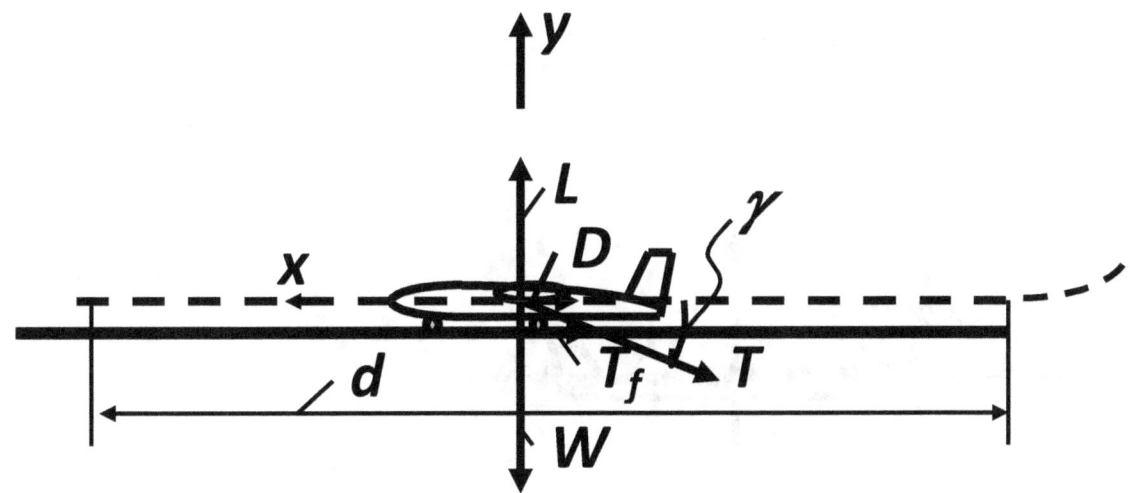

Fig. 2 Landing

The angles for take-off and landing are different, because the coefficients of rolling friction are different for take-off and landing. For take-off, the friction coefficient is small (μ approximately 0.01 – 0.05); for landing, the coefficient is larger (μ approximately 0.3 – 0.4). The direction of thrust is also different for take-off ($\gamma \cong$ +1 to +3 degrees) than for landing ($\gamma \cong$ -16 to -22 degrees). For take-off, the thrust has a forward direction; for landing the thrust has a backward direction. As a design "rule-of-thumb," we can say that the OTA in radians is equal to the coefficient of rolling friction for take-off, and the OTA is within 5% of the coefficient of rolling friction for landing. The expression tan γ = μ is exact for any rolling friction coefficient.

The optimal angles for take-off and landing are graphed in Figs. 3 and 4, respectively.

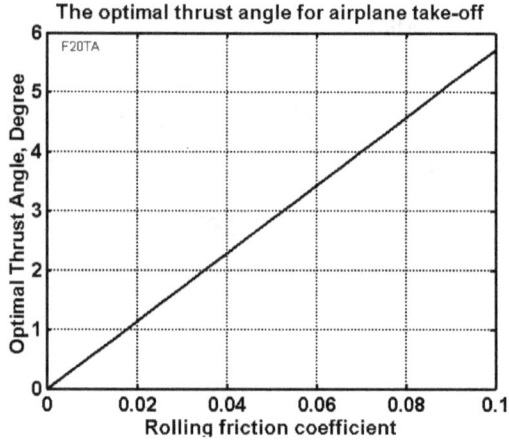

Fig. 3. Optimal Thrust Angle for Take-off

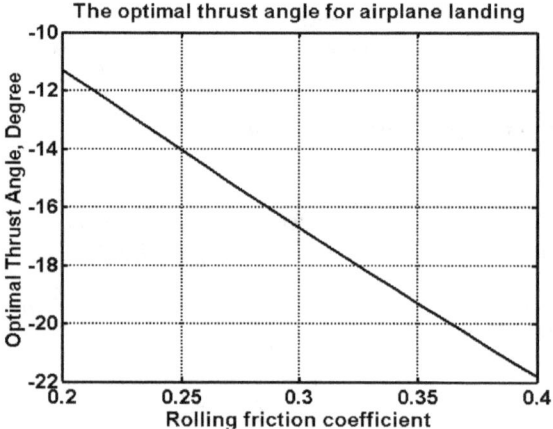

Fig.4. Optimal thrust ange for landing.

Optimal Angle of Thrust Vector in Horizontal Flight (Cruise Regime)

Assume that speed, altitude, and direction of flight are constant during horizontal flight time, and that we wish to maximize range, R, of the aircraft over the time interval $[0,T]$. Then

$$R = \int_0^T V dt \qquad (32)$$

The equilibrium equations of motion (Fig. 5) are

$$T\cos\gamma - D = 0, \qquad (33)$$

$$L - W \cdot g_0 + T\sin\gamma = 0 \qquad (34)$$

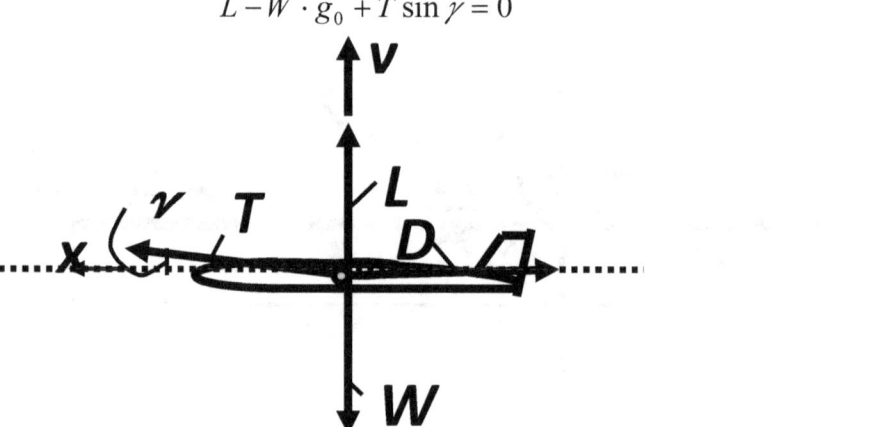

Fig. 5. Horizontal Flight.

Using the notation

$$E = \frac{L}{D} = \frac{C_L}{C_D}, \qquad u = \cos\gamma, \qquad \text{and } \sin\gamma = \sqrt{1-u^2} = 0 \qquad (35)$$

We can substitute D and u from (35) and L from (34) into (33) to obtain

$$T \cdot u - \frac{W \cdot g_0}{E} + \frac{T}{E}\sqrt{1-u^2} = 0 \qquad (36)$$

Next, compose the Hamiltonian function H, as in Eq. (10)

$$H = -V + \lambda \cdot (T \cdot u - \frac{W \cdot g_0}{E} + \frac{T}{E}\sqrt{1-u^2}) \qquad (37)$$

And find the maximum of this function

$$\frac{\partial H}{\partial u} = \lambda \cdot T\left[1 - \frac{u}{E\sqrt{1-u^2}}\right] = 0 \qquad (38)$$

If we take values for λ and T such that $\lambda \cdot T \neq 0$, we find that

$$u = E\sqrt{1-u^2} \qquad (39)$$

or

$$u^2 = E^2(1-u^2) \qquad (40)$$

From (40), it follows that

$$u = \pm \frac{E}{\sqrt{1+E^2}} \qquad (41)$$

or

$$\cos\gamma = \pm\frac{E}{\sqrt{1+E^2}} \quad \text{or} \quad \gamma = \arccos\left(\pm\frac{E}{\sqrt{1+E^2}}\right) \qquad (42)$$

Note that γ in degrees given by

$$\gamma^\circ = 180 \cdot \gamma / \pi. \qquad (43)$$

From physical conditions, it is evident that angle γ is positive. For fighter aircraft, aerodynamic efficiency, E, ranges from two to ten. For transport or passenger aircraft, efficiency ratios vary from ten to twenty. Using the trigonometric identify in (29), we get a final result in simpler form,

$$\tan\gamma = \frac{1}{E} \quad \text{or} \quad \boxed{\tan\gamma = \frac{C_D}{C_L}} \qquad (44)$$

For small γ, say γ < 0.2 radians, we have the design rule-of-thumb that

$$\boxed{\gamma \cong \frac{C_D}{C_L}} \qquad (45)$$

The OTA for the cruise regime is graphed in Fig. 6 for aerodynamic efficiencies ranging from two to ten, typical of fighter aircraft.

So far, we have used as the performance index the maximum range of the aircraft. The results are the same if we minimize fuel consumption,

$$F = \int_0^T w_f \cdot dt \qquad (46)$$

Or minimize time

$$T = \int_0^T dt \ . \qquad (47)$$

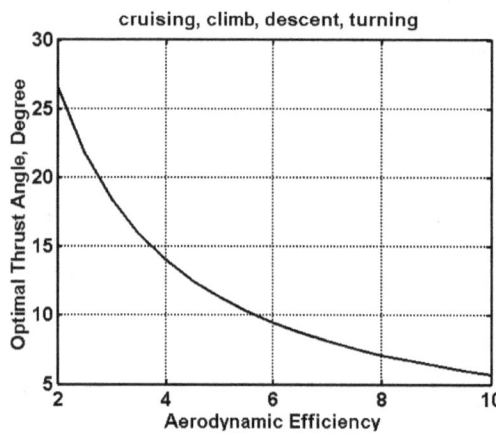

Fig. 6. Optimal Thrust Angle for Cruise Regime

Climb and Descent Regimes

Fig. 7 illustrates the climb regime.

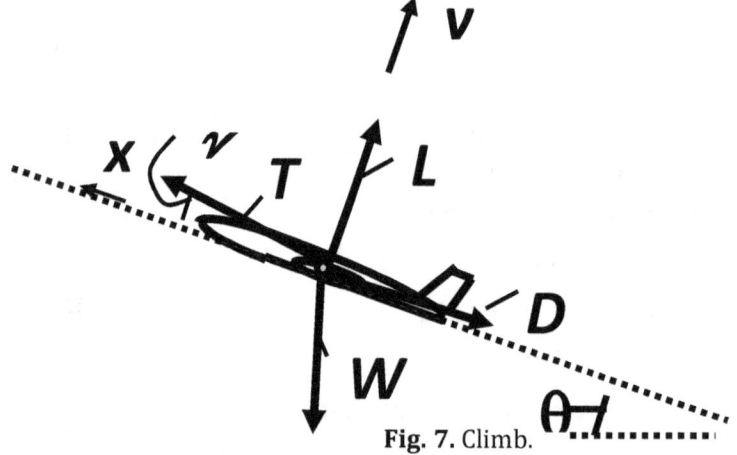

Fig. 7. Climb.

\Let us take as the performance index the range or altitude,

$$R = \int_0^T V\,dt \quad \text{or} \quad h = \int_0^T V \sin\theta\,dt \tag{48}$$

Then the equilibrium equations are

$$T\cos\gamma - D - W \cdot g_0 \sin\theta = 0 \tag{49}$$

$$L + T\sin\gamma - W \cdot g_0 \cos\theta = 0 \tag{50}$$

where θ is the angle between trajectory and horizon. Using the notation

$$E = \frac{L}{D} = \frac{C_L}{C_D}, \quad \cos\gamma = u, \quad \sin\gamma = \sqrt{1-u^2}. \tag{51}$$

and substituting Eqs. (50) and (51) into Eq. (49), we have

$$Tu - \frac{W \cdot g_0 \cos\theta}{E} + \frac{T}{E}\sqrt{1-u^2} - W\sin\theta = 0 \tag{52}$$

And

$$H = V + \lambda[Tu - \frac{W \cdot g_0 \cos\theta}{E} + \frac{T}{E}\sqrt{1-u^2} - W \cdot g_0 \sin\theta] \tag{53}$$

The necessary condition for an extreme is

$$\frac{\partial H}{\partial u} = 0 \tag{54}$$

or

$$\frac{\partial H}{\partial u} = \lambda T\left(1 - \frac{u}{E\sqrt{1-u^2}}\right) = 0 \tag{55}$$

That is the same as Eq. (38), which means that the final equation for the optimal angle of thrust vector in climb and descent will be equal to the equation for a cruise regime.

$$\cos\gamma = \pm\frac{E}{\sqrt{1+E^2}} \quad \text{or} \quad \tan\gamma = \frac{1}{E} = \frac{C_D}{C_L} \quad \text{or} \quad \gamma \approx \frac{C_D}{C_L} \tag{56}$$

From physical conditions, it is evident that angle γ is positive. The aerodynamic efficiencies are different for climb, descent, and cruise, so that the optimal thrust vector angle will be different, but the equations for the calculation are the same. Again, note that we can use trigonometric equalities to derive the more concise expression, $\cot\gamma = E$, which is exact for any aerodynamic efficiency ratio. The results are the same whether time or fuel consumption are used for the performance index.

Turning of airplane

Consider now the turning of an airplane in one plane, with a constant roll angle ϕ.
Our performance index can be distance, minimum time of turn, or fuel consumption.

$$R = \int_0^T V dt, \qquad T = \int_0^T dt, \qquad F = \int_0^T w_f dt \qquad (57)$$

The equations of motion are

$$T\cos - D = 0 \qquad (58)$$

$$L + T\sin\gamma - W \cdot g_0 \cos\phi = 0 \qquad (59)$$

Using the notation

$$E = \frac{L}{D} = \frac{C_L}{C_D}; \qquad \cos\gamma = u; \qquad \sin\gamma = \sqrt{1-u^2} \qquad (60)$$

and substituting (59) and (60) into (58), we get

$$T \cdot u - \frac{W \cdot g_0 \cos\phi}{E} + \frac{T\sqrt{1-u^2}}{E} = 0 \qquad (61)$$

and

$$H = V + \lambda\left[Tu - \frac{W \cdot g_0 \cos\phi}{E} + \frac{T}{E}\sqrt{1-u^2}\right] \qquad (62)$$

The necessary condition for an extreme is

$$\frac{\partial H}{\partial u} = 0 \quad \text{or} \quad \frac{\partial H}{\partial u} = \lambda T\left(1 - \frac{u}{E\sqrt{1-u^2}}\right) = 0 \qquad (63)$$

Eq. (63) is equivalent to Eq. (38), which means the final equation for the optimal angle of thrust vector in a roll is equal to the equation for a cruise regime

$$\cos\gamma = \pm\frac{E}{\sqrt{1+E^2}} \quad \text{or} \quad \tan\gamma = \frac{1}{E} = \frac{C_D}{C_L} \quad \text{or} \quad \gamma \approx \frac{C_D}{C_L} \qquad (64)$$

From physical conditions, it is evident that angle γ is positive.

Discussion

The problem of determining an OTA is also discussed in Reference 9, in which the OTA for rocket-powered aircraft is given by

$$\omega = \arctan \frac{\partial D}{\partial L} \qquad (65)$$

where ω is equivalent to our angle γ. In the particular case of a parabolic polar drag coefficient of the form $C_D = C_{D_0}(M) + K(M)C_L^2$, where M is the Mach number, K is the induced drag factor, and C_{D_0} is the zero-lift drag coefficient, Eq. (65) leads to

$$\omega = \arctan(2KC_L) \qquad (66)$$

Eqs. (44) and (66) give very different results (Fig. 8). For example, when there is no lift force ($C_L = 0$), Eq. (44) gives γ = 90°, meaning that the optimal thrust angle is strictly vertical (perpendicular to the desired trajectory), while Eq. (66) gives ω = 0, corresponding to a horizontal thrust. Conversely, when the lift force is maximum, Eq. (65) gives ω = 90°. We also see in Fig. 8 that as the lift force (C_L) decreases after passing through its maximum point, Eq. (65) yields an optimal thrust angle greater than 90°, producing a reverse thrust force. So, we conclude that Eqs. (65) and (66) do not adequately model the OTA near extreme points.

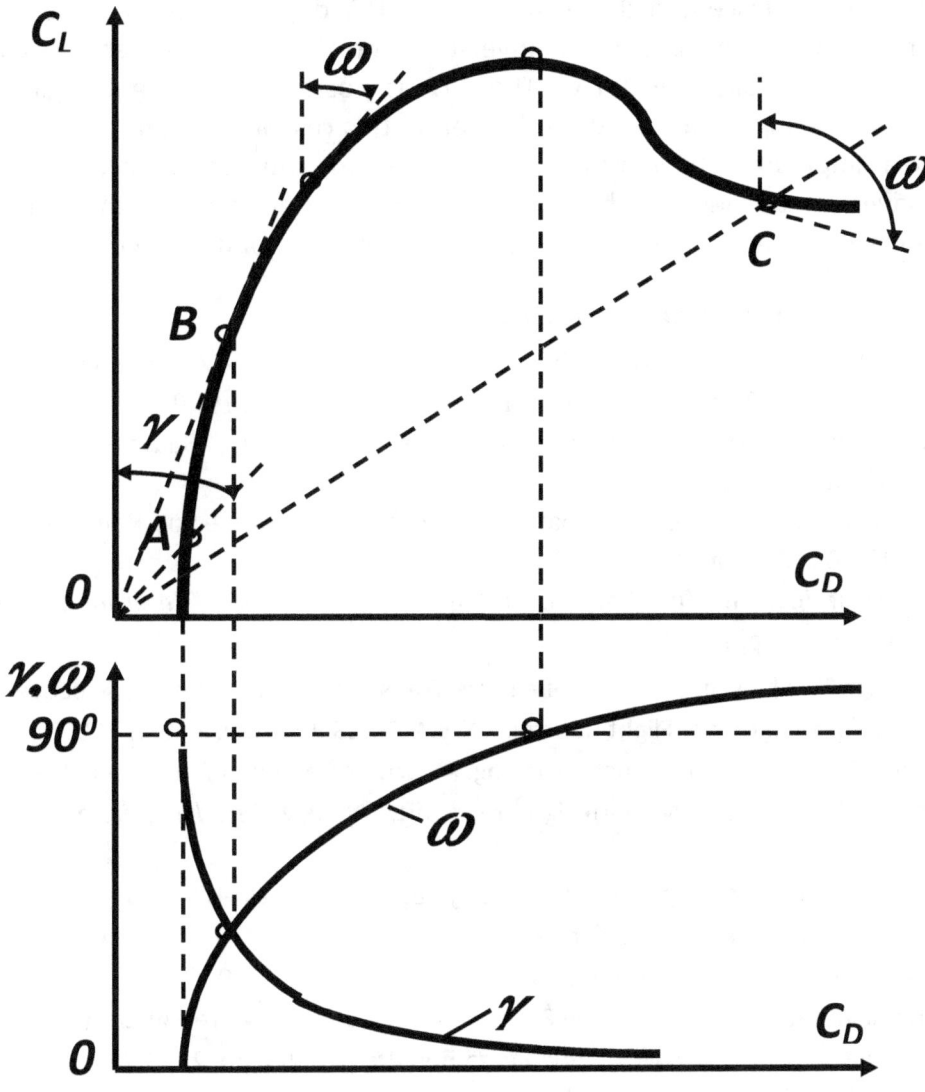

Fig. 8. Comparison of γ and ω.

The angle γ produced by Eq. (44) also has better trend characteristics, starting at 90° when $C_L = 0$, then decreasing as the lift force is increasing and positive. The angle ω starts at zero when $C_L = 0$, then increases as the aerodynamic lift force increases. Eqs. (44) and (65) do produce the same result at one point, when the efficiency coefficient, $E = C_L / C_D$ is maximized. Eq. (65) was derived for an optimal angle of attack, and the result is valid at the point of optimal aircraft lift. Eq. (44) is more general, and may be used at any polar coordinate in any of the four flight regimes: climb, cruise, turn or descent.

Conclusions

In this chapter, we derived two simple equations for the optimal thrust angle of an aircraft. One equation is valid for take-off and landing, the other for climb, cruise, turn, and descent. During take-off, the OTA is positive, decreases as the coefficient of rolling friction decreases, and is essentially equal to the friction coefficient. During landing, the OTA is negative, increases as the coefficient of rolling friction increases, and is within five percent of the value of the friction coefficient. The simple expression tan OTA = μ provides an exact result for the OTA as a function of the rolling friction coefficient.

In the climb, cruise, turn or descent flight regimes, the OTA depends only on the coefficient of aerodynamic efficiency. Here we observe an inverse proportion: the greater the coefficient of aerodynamic efficiency, the smaller the OTA. The OTA is positive in all flight regimes, with the possible exception of air braking, which is not addressed in this research. As in the cases of take-off and landing, we have a simple expression, tan OTA = 1 / E, relating the optimal thrust angle to a single parameter, the aerodynamic efficiency. The equations for OTA developed in this paper were also shown to have more intuitive trends and better behavior at extreme points than the Miele equations.

References to Chapter 12

[1] Bolonkin, Alexander, "A New Approach to Finding a Global Optimum," *New American's Collected Scientific Reports*, Vol. 1, The Bnai Zion Soviet-American Scientists Division, 1990.

[2] Bolonkin, Alexander, "New Methods of Optimization and their Application," Bauman Technical University, Moscow, 1972 (in Russian).|

[3] Gilyard, Glenn and Bolonkin, Alexander, "Optimal Pitch Thrust-Vector Angle and Benefits for all Flight Regimes, NASA-TM-2000-209021, March 2000.

[4] Gal-Or, Benjamin, "Thrust Vectoring for Flight Control and Safety: A Review," *International Journal of Turbo and Jet Engines*, 11, 1994, pp. 119-136.

[5] Gerren, Donna S., "Design, Analysis, and Control of a Large Transport Aircraft Utilizing Selective Engine Thrust as a Backup System for the Primary Flight Control." NASA-CR-186035, September 1995.

[6] Mangold, P., and Wedekind, G, "Inflight Thrust Vectoring: A Further Degree of Freedom in the Aerodynamic/Flight Mechanical Design of Modern Fighter Aircraft," *AGARD, Aerodynamics of Combat Aircraft Controls and of Ground Effects*, April 1990.

[7] Koelle, Heinz Hermann, Ed., *Handbook of Astronautical Engineering*, McGraw-Hill, 1961, pp. 19-38.

[8] Kutschera, Antony, and Render Peter M., "Performance assessment of thrust vector controlled post stall manoeuvrable fighter aircraft using minimal input data," AIAA Paper 99-4020, 1999.

[9] Miele, Angelo, "General Variational Theory of the Flight Paths of Rocket-Powered Aircraft, Missiles and Satellite Carriers." *Astronautica acta*, Vol 4, New York: Pergamon Press, 1958, pp. 272-273.

[10] Pontryagin, L.S., V.G. Boltyanskii, R.V. Gamkrelidze, and E.F. Mischenko, *The Mathematical Theory of Optimal Processes*. New York: Interscience Publishers, Inc., 1962, pp. 17 - 20.

Chapter 13.
Optimal Trajectories of Aerospace Vehicles*

Summary

The author developed a theory of optimal trajectories for air vehicles with variable wing area and with conventional wings. He applied a new theory of singular optimal solutions and obtained in many cases the optimal flight. The wing drag of a variable area wing does not depend on air speed and air density. At first glance the results may seem strange however, this is correct and this paper will show how this new theory may be used. The equations that follow allow computing the optimal control and optimal trajectories of subsonic aircraft with pistons, jets, and rocket engines, supersonic aircraft, winged bombs with and without engines, hypersonic warheads, and missiles with wings.

The main idea of the research is in using the vehicle's kinetic energy for increasing the range of missiles and projectiles.

The author shows that the range of a ballistic warhead can be increased 3-4 times if an optimal wing is added to the ballistic warhead, especially a wing with variable area. If we do not need increased range, the warhead mass can be increased. The range of big gun shells can also be increased 3 - 9 times. The range of aircraft may be improved 3-15% and more.

The results can be used for the design of aircraft, missiles, flying bombs and shells of big guns.

Key words: optimal trajectory, singular optimal solution, range, aircraft, missiles, and projectiles.

*Theory presented to AIAA/NASA/USAF/SSMO Symposium on Multidisciplinary Analysis and Optimization, Panama City, Florida, USA, Sept. 7-9, 1994. Full text is published in AEAT, Vol.76, No.2, 2004, pp.193-214.

Nomenclature (in metric system)

a = the sound speed,
a_1, b_1, a_2, b_2 are coefficients of an exponential atmosphere,
C_L = lift coefficient,
C_D = drag coefficient,
C_{Do} = drag coefficient for $C_L=0$,
C_{DW} = wave wing drag coefficient when $\alpha = 0$,
C_{Db} body drag coefficient,
c = relative thickness of a wing,
c_b is relative thickness of body,
c_1 = relative thickness of a vehicle body,
c_s = fuel consumption, kg/sec/ kg trust,
$\overline{D}$ = drag of vehicle,
D = drag of vehicle without α,
D_{0W} = wave wing drag when $\alpha = 0$,
D_{0b} = drag of a vehicle body,
H = Hamiltonian,
h = altitude,
$K=C_L/C_D$ is the wing efficiency coefficient,
k_1, k_2, k_3 are vehicle average aerodynamic efficiencies for 1, 2, 3 sub distances respectively,
L = range,

$M=V/a$ is Mach number,
m = mass of vehicle,
$p=m/S$ is a load on a wing square meter,
$q=\rho V^2/2$ is a dynamic air pressure,
R is aircraft range or R = distance from flight vehicle to Earth center; $R = R_o+h$, where R_o=6378 km is Earth radius,
t = time,
$T = V_e \beta$ is thrust,
V = vehicle speed,
V_e = speed of throw back mass (air for propeller engine, jet for jet and rocket engine),
S = wing area,
s = length of trajectory,
T = engine trust,
Y = lift force,
α = wing attack angle,
β = fuel consumption,
θ = angle between the vehicle velocity and horizon,
ω = thrust angle between a thrust and velocity,
ω_E = Earth angle speed,
φ_E = lesser angle between Earth Pole axis and perpendicular to a flight plate,
ρ = air density.

Introduction

The topic of the optimal flight of air vehicles is very important. There are numerous articles and books about the optimal trajectories of rockets, missiles, and aircraft. The classical research of this topic is in [1]. Unfortunately, the optimal theory of this problem is very complex. In most cases, the researchers obtained the complex equations, which allow one to compute a single optimal trajectory for a given aircraft and for given conditions, but the structure of optimal flight is not clear and simple formulas of optimal control (which depends only on flight conditions) are absent.

The author's new theory of singular optimal solutions, developed earlier in [2]-[10], does not contain unknown coefficients or variables as previous theories have. He found that optimal flight path depend only on flight conditions and the addition of certain variable wing structures.

In conclusion, the author applies his solution to ballistic missiles, warheads, flying bombs, big gun shells, and subsonic, supersonic, and hypersonic aircraft with rocket, turbo-jet, and propeller engines. He shows that the range of these air vehicles can be increased 3-9 times.

1. General equations

Let us consider the movement of an air vehicle with the following conditions:
  1. The vehicle moves in a plane containing the Earth's center. 2. The vehicle design alloys to change the wing area (this will prove important in the remainder of this article). 3. We neglect the centrifugal force from Earth rotation (it is less 1%). 4. Earth has a curvature.

Then the equation of flight vehicle (in system coordinate when a center system is located in a center of gravity of the flight vehicle, the axis x - in a flight direction, the axis y - in a perpendicular direction of the axis x, fig.1-1) are

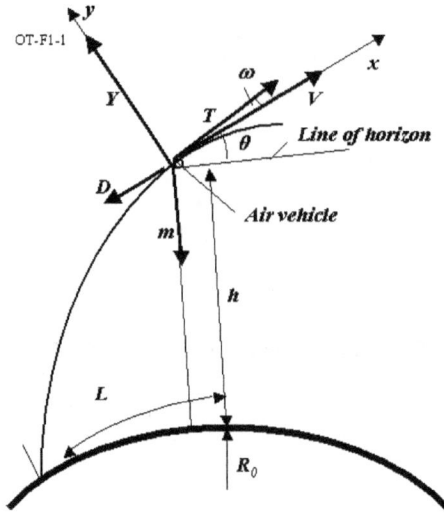

Fig.1-1. Vehicle forces and coordinate system.

$$\frac{dL}{dt} = V\cos\theta ,$$
$$\frac{dh}{dt} = V\sin\theta ,$$
(1-1)-(1-2)

$$\frac{dV}{dt} = \frac{T(h,V,\beta)\cos\omega - \overline{D}(\alpha,V,h)}{m} - g\sin\theta ,$$
$$\frac{d\theta}{dt} = \frac{T(h,V,\beta)\sin\omega + Y(\alpha,V,h)}{mV} - \frac{g}{V}\cos\theta + \frac{V\cos\theta}{R} + 2\omega_E\cos\varphi_E ,$$
$$\frac{dm}{dt} = -\beta ,$$
(1-3)-(1-5)

All values are taken in metric system and all angles are taken in radians.

2. Flight with small change of vehicle mass and flight path angle

Most air vehicles fly with angle θ in the range $\pm 15^0$ ($\theta = \pm 0.2618$ rad) and the engine located along the velocity vector. It means

$$\sin\theta = \theta, \qquad \cos\theta = 1, \qquad \omega = 0,$$
(2-1)-(2-3)

because $\sin 15^0 = 0.25882$, $\cos 15^0 = 0.9659$.

Let as substitute (2-1)-(2-3) in (1-1)-(1-5)

$$\frac{dL}{dt} = V ,$$
$$\frac{dh}{dt} = V\theta ,$$
(2-4)-(2-5)

$$\frac{dV}{dt} = \frac{T(h,V) - \overline{D}(\alpha,V,h)}{m} - g\theta,$$

$$\frac{d\theta}{dt} = \frac{Y(\alpha,V,h)}{mV} - \frac{g}{V} + \frac{V}{R} + 2\omega_E \cos\varphi_E,$$

(2-6)-(2-8)

$$\frac{dm}{dt} = -\beta,$$

(2-9)

where

$$|\theta| \le \theta_{max}.$$

(2-10)

A lot of air vehicles fly with the small angular speed $d\theta/dt$. The change of mass is also small in flight. It means m = const, $dm/dt \cong 0$.

$$d\theta/dt \approx 0, \quad dm/dt = 0.$$

(2-11)- (2-12)

Let us take a new independent variable s = length of trajectory

$$dt = ds/V,$$

(2-13)

and substitute (2-10)-(2-13) in (2-4)-(2-9). Then system (2-4)-(2-9) takes the form

$$\frac{dL}{ds} = 1,$$

$$\frac{dh}{ds} = \theta,$$

$$\frac{dV}{ds} = \frac{T(h,V) - \overline{D}(\alpha,V,h)}{mV} - \frac{g}{V}\theta,$$

$$0 = \frac{Y(\alpha,V,h)}{mV} - \frac{g}{V} + \frac{V}{R} + 2\omega_E \cos\varphi_E.$$

(2-14)-(2-17)

Let us to re-write the equation (2-17) in form

$$Y(\alpha,V,h) - mg + \frac{mV^2}{R} + 2mV\omega_E \cos\varphi_E = 0.$$

(2-18)

If we neglect the last member, the equation (2-18) takes form

$$Y(\alpha,V,h) - mg + \frac{mV^2}{R} = 0.$$

(2-18)'

If the V is not very large (V< 3 km/sec), the two last members in the equation (2-17) is small and they may be neglected. The equations (2-18), (2-18)' can be used for deleting α from $\overline{D}$.

Note the new drag without α as

$$D=D(h,V).$$

(2-19)

If we substitute α from (2-18) to Eq. (2-16) the equation system take the form

$$\frac{dL}{ds} = 1,$$

$$\frac{dh}{ds} = \theta,$$ (2-20)-(2-22)

$$\frac{dV}{ds} = \frac{T(h,V) - D(V,h)}{mV} - \frac{g}{V}\theta,$$

Here the variable θ is new control limited by

$$|\theta| \leq \theta_{max}.$$ (2-23)

Statement of problem.

Consider the problem: find the maximum range of an air vehicle described by equation (2-20) - (2-22) for limitation (2-23). For readers are not friendly with Method of Deformation, we can also solve this problem by a conventional method using the Hamiltonian for this problem (as it is shown in Chapter 2 most conventional methods may be received from Method of Deformation). However, this is a non-linear problem and contains the linear control. That means, this problem has a singular solution. For finding the singular solution, we will use the methods developed in [2] and [4].

Write the Hamiltonian

$$H = 1 + \lambda_1 \theta + \lambda_2 \frac{1}{V}\left(\frac{T-D}{m} - g\theta\right),$$ (2-24)

where $\lambda_1(s)$, $\lambda_2(s)$ are unknown multipliers. Application of the conventional method gives

$$\dot{\lambda}_1 = -\frac{\partial H}{\partial h} = -\lambda_2 \frac{1}{V}\left(\frac{T'_h - D'_h}{m}\right),$$

$$\dot{\lambda}_2 = -\frac{\partial H}{\partial V} = -\lambda_2\left[-\frac{1}{V^2}\left(\frac{T-D}{m} - g\theta\right) + \frac{1}{V}\left(\frac{T'_V - D'_V}{m}\right)\right],$$ (2-25)-(2-27)

$$\theta = \max_\theta H = \theta_{max} sign\left[\lambda_1 - \lambda_2 \frac{g}{V}\right].$$

Where D'_h, D'_v, T'_h, T'_v denote the first partial derivatives D, T per h, V respectively.

The last equation shows that control θ can have only two values $\pm\theta_{max}$. We consider the singular case when

$$A = \lambda_1 - \lambda_2 \frac{g}{V} \equiv 0.$$ (2-28)

This equation has two unknown variables λ_1 and λ_2 and does not contain information about the control θ.

Let us, as in [2], to differentiate equation (2-28) to the independence variable s. After substitution the equations (2-22), (2-25), (2-26), (2-28), we received the relation for $\lambda_1 \neq 0$, $\lambda_2 \neq 0$

$$\boxed{V(T'_h - D'_h) = g(T'_V - D'_V)}$$ (2-29)

This equation does not contain θ either, but it contains the important relation between the variables V and h on the optimal trajectory.

If we know formulas (or graphs)

$$D = D(h,V), \qquad (2\text{-}30)$$

$$T = T(h,V), \qquad (2\text{-}31)$$

we could find the relation

$$h = h(V) \qquad (2\text{-}32)$$

and the optimal trajectory for a given air vehicle.

That also gives the important information about the structure of the optimal solution. The investigation of the equation (2-29) shows that the equation has one solution in each, the subsonic, supersonic, and hypersonic fields. The equation can have two solutions for a transonic field.

It means the optimal trajectory in most cases has three parts (see fig. 2-1):

a) In a climb and flight: a vehicle moves from the initial point A with the angle $\pm\theta_{max}$ up to the optimal curve (2-32), then one moves up along the optimal curve (2-32), and further that moves with the angle $\pm\theta_{max}$ to the point B.

b) In a descent and flight (Fig.2-2): a vehicle moves from the initial point A with the angle $\pm\theta_{max}$ (up or down) to the optimal curve (2-32), then one moves down along the optimal curve (2-32), and further that moves with the angle $\pm\theta_{max}$ (up or down) to the point B.

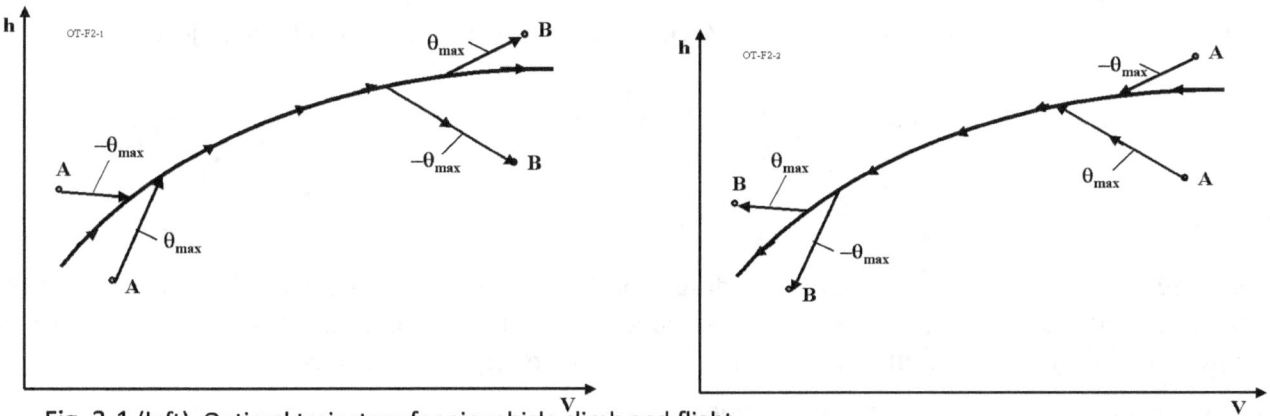

Fig. 2-1 (left). Optimal trajectory for air vehicle climb and flight.

Fig.2-2 (right). Optimal trajectory for air vehicle descent and flight.

The selection of direction (up or down, with θ_{max} or $-\theta_{max}$ respectively) depends only on the position of the initial and end points A and B.

For air vehicles with rocket engines T=const, the equation (2-29) has very simple form

$$\boxed{VD'_h = gD'_V} . \qquad (2\text{-}33)$$

The same form (same curve) is also for a ballistic warhead, which does not have engine thrust (after it's short initial burn) (T=0).

If we want to have an equation for the control θ, we continue to differentiate equation (2-29) with the independent variable s, and substitute the equations (2-21), (2-22), (2-25), (2-26), (2-28), (2-29). We received the relation for θ if $\lambda_1 \neq 0$, $\lambda_2 \neq 0$

$$\theta = \frac{B_1(T-D)}{mV\left(B_1\frac{g}{V} - B_2\right)}, \qquad (2\text{-}34)$$

where

$$\begin{aligned}B_1 &= (T_h' - D_h') + V(T_{hV}'' - D_{hV}'') - g(T_{VV}'' - D_{VV}''), \\ B_2 &= V(T_{hh}'' - D_{hh}'') - g(T_{hV}'' - D_{hV}'').\end{aligned} \qquad (35)\text{-}(36)$$

Here signs in form D_{hV}'' are the second partial derivates D for h, V.

$$D_{hV}'' = \frac{\partial^2 D}{\partial h \partial V}. \qquad (2\text{-}37)$$

If the thrust does not depend from h, V (T=const) or no engine (T = 0), the equation for θ became simpler

$$\theta = \frac{[(gD_{VV}'' - D_h') - VD_{hV}''](T-D)}{m[g(gD_{VV}'' - D_h') + V^2 D_{hh}'']}. \qquad (2\text{-}38)$$

With according [2]-[8] (see, for example, [4], Eq. (4.2)) the necessary condition of optimal trajectory is

$$-(-1)^k \frac{\partial}{\partial \theta}\left[\frac{d^{2k}}{ds^{2k}}\left(\frac{\partial H}{\partial \theta}\right)\right] \geq 0. \qquad (2\text{-}39)$$

where k = 1.

For getting results for different forms of the drags and thrusts, we must take formulas (or graphs) for subsonic, transonic, supersonic, hypersonic speed, specific formulas for the thrust and substitute them in the equation (2-29), (2-34). Consider two cases: subsonic and hypersonic speeds.

Subsonic speed (V<270 m/sec) and different engines

Lift, drag, and derivative equations for subsonic speed are

$$L = mg = \varsigma\alpha\frac{\rho V^2}{2}S, \quad \overline{D} = C_D\frac{\rho V^2}{2}S, \quad C_D = C_{Do} + \varepsilon\alpha^2, \quad D = \left[C_{Do} + \varepsilon\left(\frac{2mg}{\varsigma\rho V^2 S}\right)^2\right]\frac{\rho V^2}{2}S,$$

$$\rho = a_1 e^{-h/b_1}, \quad \frac{\partial D}{\partial V} = \left[C_{Do} - \varepsilon\left(\frac{2mg}{\varsigma\rho V^2 S}\right)^2\right]\rho VS, \quad \frac{\partial D}{\partial h} = -\frac{1}{b_1}\left[C_{Do} - \varepsilon\left(\frac{2mg}{\varsigma\rho V^2 S}\right)^2\right]\frac{\rho V^2}{2}S, (2-40)$$

where. $\varsigma = \dfrac{6.24\lambda}{\lambda+2}$, $\varepsilon = \dfrac{\varsigma^2}{\pi\lambda}$, Magnitude $\varepsilon \approx \varsigma^2/\pi\lambda$ is an induced drag coefficient, $\lambda = l^2/S$, l is a wing span.

It is known in conventional aerodynamics that the coefficient of the flight efficiency k is

$$k = \frac{C_L}{C_D} = \frac{\varsigma\alpha}{C_{Do} + \varepsilon\alpha^2}, \quad from \quad \max_\alpha k \quad we\ get \quad \alpha_{opt} = \sqrt{\frac{C_{Do}}{\varepsilon}}, \quad k_{max} = \frac{\varsigma}{2\sqrt{\varepsilon C_{Do}}}, \tag{2-41}$$

a) Aircraft with rocket engine. For this aircraft the thrust T is constant or 0. The equation (2-29) has form (2-33). Find the partial derivatives

$$T'_V = 0, \quad T'_h = 0. \tag{2-42}$$

Substitute (2-40)-(2-42) in (2-33) we get the relation between an air density ρ, altitude h, and aircraft speed V:

$$\rho = \frac{2gp}{\varsigma V^2}\sqrt{\frac{\varepsilon}{C_{Do}}}, \quad p = \frac{m}{S}, \quad h = b_1 \ln\frac{a_1}{\rho}, \tag{2-43}$$

where $p = m/S$ is a load on a wing meter square. For diapason $h = 0 \div 11$ km the coefficients $a_1 = 1.225$, $b_1 = 9086$.

Results of the computation is presented in fig. 2-3.

b) Aircraft with turbo-jet engine. The trust for this engine is

$$T = T_0\frac{\rho}{\rho_0}, \quad T'_h = -\frac{T}{b_1}, \quad T'_V = 0. \tag{2-44}$$

Substitute (2-44) in (2-29). We get

$$V\left(-\frac{T}{b_1} - D'_h\right) = -gD'_v \quad or \quad T = \frac{b_1}{V}(gD'_V - VD'_h), \tag{2-44'}$$

substitute (2-40), (2-44) in (2-29), we get

$$\frac{1}{p}\left(\frac{V^2}{2b_1} + g\right)\left[C_{Do} - \varepsilon\left(\frac{2pg}{\varsigma\rho V^2}\right)^2\right] = \frac{\overline{T}_0}{b_1\rho_0}, \quad where \quad \overline{T}_0 = \frac{T_0}{m}. \tag{2-45}$$

Find ρ, h from (2-45)

$$\rho = \frac{2pg\sqrt{\varepsilon}}{\varsigma V^2 \sqrt{A_2}}, \quad where \quad A_2 = C_{Do} - \frac{2p\overline{T}_0}{\rho_0(V^2 + 2b_1 g)} \quad \overline{T}_0 = \frac{T_0}{m}, \quad h = b_1 \ln\frac{a_1}{\rho}. \tag{2-46}$$

c) Result of computation for the different p, $T = 0.8$ N/kg, $a_1 = 1.225$, $b_1 = 9086$ are presented in fig. 2-4.

Piston and turbo engines with propeller. All current propeller engines have propellers with variable pitch. Propeller coefficient efficiency, η, approximately is constant. The trust of this engine is

$$T = \frac{N_0}{V}\frac{\rho}{\rho_0}, \quad T'_V = -\frac{T}{V}, \quad T'_h = -\frac{T}{b_1}., \tag{2-47}$$

where $N_0 = N_e \eta$, N_e = engine power at $h = 0$.

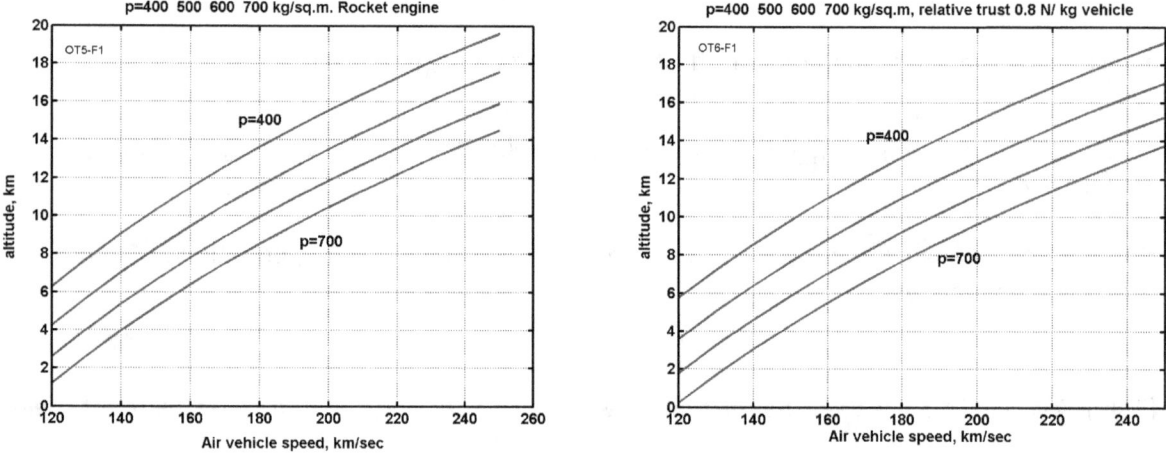

Fig.2-3 (left). Air vehicle altitude versus speed for the wing load p=400, 500, 600, 700 kg/m² and rocket engine.

Fig.2-4 (right). Air vehicle altitude versus speed for wing load p=400, 500, 600, 700 kg/m², turbo-jet engine, and relative trust 0.8 N/kg vehicle.

Substitute (2-47) in (2-29). We get the equation for trust

$$V\left(\frac{T}{b_1} + D_h'\right) = g\left(\frac{T}{V} + D_V'\right) \quad or \quad T = \frac{b_1 V(gD_V' - VD_h')}{V^2 - gb_1} . \qquad (2\text{-}47)'$$

Substitute (2-40), (2-47) in (2-29). We get

$$\frac{V}{p}\left(\frac{V^2}{b_1} - g\right)\left[C_{Do} - \varepsilon\left(\frac{2pg}{\rho V^2}\right)^2\right] = \frac{\overline{N}_0}{\rho_0}\left(\frac{g}{V^2} - \frac{1}{b_1}\right), \quad where \quad \overline{N}_0 = \frac{N_0}{m}, \quad p = \frac{m}{S} . \qquad (2\text{-}48)$$

Find ρ, h from (2-48)

$$\rho = \frac{2pg\sqrt{\varepsilon}}{\varsigma V^2 \sqrt{A_3}}, \quad where \quad A_3 = C_{Do} + \frac{p\overline{N}_0}{\rho_0 V^3}, \quad h = b_1 \ln\frac{a_1}{\rho} . \qquad (2\text{-}49)$$

Result of computation for C_{Do}=0.025, λ=10, different p, N are presented in fig. 2-5.

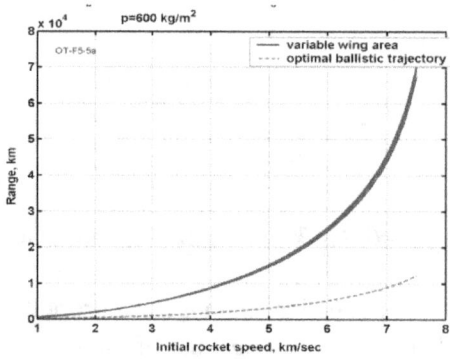

Fig.2-5. Air vehicle altitude versus speed for the wing load p = 250, 300, 350, 400 kg/m², piston (propeller) engine, and relative engine power 100 W/kg vehicle.

Hypersonic speed (1 km/s< V<7 km/s).

The lift and drag forces in hypersonic flight approximately equal (see (2-18'))

$$L(\alpha,V,h) = mg - \frac{mV^2}{R} = \varsigma\alpha\frac{a\rho V}{2}S, \quad \overline{D} = (C_{DW} + \varepsilon\alpha^2)\frac{a\rho V}{2}S + C_{Db}\frac{a\rho V}{2}S_b,$$

$$\alpha = \frac{2p(g-V^2/R)}{\varsigma\rho aV}, \quad D = \left[C_{DW}\frac{a\rho V}{2} + \frac{2\varepsilon}{\rho aV}\left(\frac{m(g-V^2/R)}{\varsigma S}\right)^2\right]S + C_{Db}\frac{a\rho V}{2}S_b, \qquad (2\text{-}50)$$

or $\quad \dfrac{D}{m} = \left[C_{DW}\dfrac{q}{p} + \dfrac{\varepsilon p}{q}\left(\dfrac{g-V^2/R}{\varsigma}\right)^2\right] + C_{Db}\dfrac{q}{p_b}, \quad q = \dfrac{\rho aV}{2},$

Note

$$D_{0W} = C_{DW}\frac{\rho aV}{2}S, \quad D_{0b} = C_{Db}\frac{\rho aV}{2}S_b, \quad C_{DW} = 4c, \quad C_{Db} = 2c_b, \quad \rho = a_2 e^{-\frac{h-11000}{b_2}}, \qquad (2\text{-}51)$$

The derivatives of D by V, h are

$$D'_V = \frac{D_{0W}}{V} + \frac{D_{0b}}{V} - \frac{2\varepsilon mp}{\varsigma^2\rho a}\left(g-\frac{V^2}{R}\right)\left(\frac{3}{R}+\frac{g}{V^2}\right),$$

$$D'_h = D'_\rho \rho'_h = -\frac{1}{b_2}\left(D_{0W} + D_{0b} - \frac{2\varepsilon mp(g-V^2/R)^2}{\varsigma^2\rho aV}\right) \qquad (2\text{-}52)$$

a) Rocket engine or hypersonic glider. The derivatives from $T=const$ and $T=0$ are

$$T'_V = 0, \quad T'_h = 0. \qquad (2\text{-}53)$$

Substitute (2-51) in (2-52), expressions (2-52), (2-53) in (2-33) and find ρ, h. We get for $h > 11,000$ m

$$\rho = \frac{2p\sqrt{\varepsilon}}{\varsigma a}\sqrt{A_4}, \quad A_4 = \frac{\left(g-\dfrac{V^2}{R}\right)\left[g\left(\dfrac{3}{R}+\dfrac{g}{V^2}\right)+\dfrac{1}{b_2}\left(g-\dfrac{V^2}{R}\right)\right]}{\left(\dfrac{V^2}{b_2}+g\right)\left[C_{DW}+C_{Db}\left(\dfrac{S_b}{S}\right)\right]}, \quad h = 11000 + b_2\ln\frac{a_2}{\rho}, \qquad (2\text{-}54)$$

where $a_2 = 0.365$, $b_2 = 6997$ are coefficients of the exponent atmosphere for stratosphere 11 to 60 km.

If we neglect the small member $g\left(\dfrac{3}{R}+\dfrac{g}{V^2}\right)$ for $M > 3$ in (2-54), the equations get a form

$$\rho = \frac{2p(g-V^2/R)\sqrt{\varepsilon}}{\varsigma a}\sqrt{A_5}, \quad A_5 = \frac{1}{C_{Do}(V^2+gb_2)}, \quad \text{where} \quad C_{Do} = C_{0W} + C_{Db}\left(\frac{S_b}{S}\right),$$

where $C_{DW} \approx 4c$. If we neglect the member gb_2 (for $M > 3$), then

$$\rho = \frac{2p(g - V^2/R)}{\varsigma aV} \sqrt{\frac{\varepsilon}{C_{Do}}} \; . \tag{2-55}$$

In the limit as $R \to \infty$ in (2-54), we find

$$\rho = \frac{2pg}{\varsigma aV} \sqrt{\frac{\varepsilon}{C_{Do}}} \; . \tag{2-55}'$$

Here $\sqrt{C_{Do}/\varepsilon} = \alpha_{opt}$ is an optimal (maximum C_L/C_D) wing attack angle of the horizontal flight.

Results of computation (2-54) are presented in fig. 2-6.

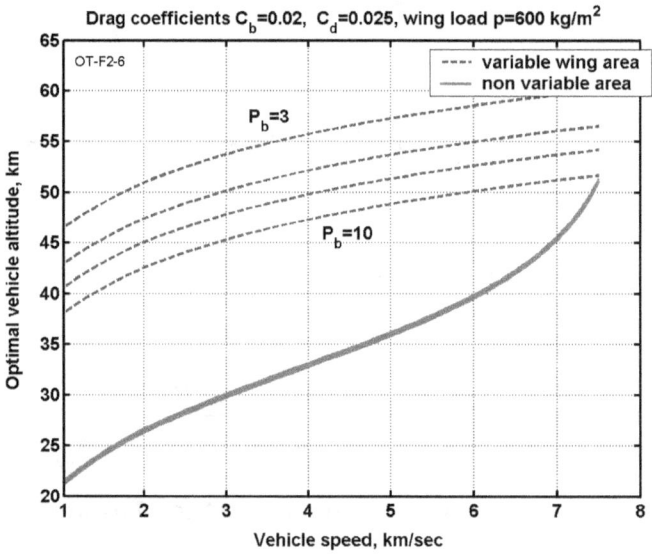

Fig. 2-6. Optimal vehicle altitude versus speed for specific body load P_b=3, 5, 7, 10 ton/m², body drag coefficient C_b=0.02, wing drag coefficient C_d= 0.025, wing load p = 600 kg/m².

d) Ramjet engine. The trust of jet engine approximately equals ($M < 4$)

$$T = \xi \frac{\rho}{\rho_2} V^2, \quad T_V' = \frac{2T}{V}, \quad T_h' = -\frac{T}{b_2}, \tag{2-56}$$

where ξ is a numerical coefficient, ρ_2 is the air density in the lower end of the selected atmospheric diapason (in our case 11 km).

Substitute (2-56), (2-52) in our main equation (2-29). By repeat reasoning we can get the equation for the given engine

$$\rho = \frac{2p\sqrt{\varepsilon}}{\varsigma a}\sqrt{A_6}, \quad A_6 = \frac{(g - V^2/g)\left[g\left(\frac{3}{R} + \frac{g}{V^2}\right) + \frac{1}{b_2}(g - V^2/R)\right]}{\left[C_{DW} + C_{Db}\left(\frac{S_b}{S}\right)\right]\left[\left(\frac{V^2}{b_2} + g\right) - \frac{2\overline{T}_0 p}{a\rho_0}\left(\frac{V}{b_2} - \frac{2g}{V}\right)\right]}, \quad \overline{T} = \frac{T_0}{m}, \tag{2-57}$$

where T_0 is taken in the lower end of the exponent atmospheric diapason (in our case 11 km). The curve of the air density vs. the altitude h is computed similar to (2-54).

Optimal wing area

The lift force and drag of any wing may be written as

$$Y = mg = Y(\alpha, q, S), \quad D = D(\alpha^2, q, S). \tag{2-58}$$

Substitute (2-58) in (2-24) and find minimum H vs. S, we get equation

$$D + \overline{D}'_\alpha \alpha'_S S = 0, \quad \text{or} \quad D + D'_S S = 0, \tag{2-59}$$

where α is value found from the first equation (2-58). The equation (2-59) is the general equation of the optimal wing area and optimal specific load $p=m/S$ on a wing area.

a) **Subsonic speed.** Lift force and drag of the subsonic wing are

$$Y = mg = \varsigma \alpha q S \quad \text{or} \quad \alpha = \frac{mg}{\varsigma q S}, \quad \overline{D} = (C_{Do} + \varepsilon \alpha^2) q S \quad D = C_{DW} q S + \varepsilon \left(\frac{mg}{\varsigma}\right)^2 \frac{1}{qS}, \tag{2-58}'$$

where $q = \rho V^2/2$ is a dynamic air pressure for the subsonic speed.

Substitute the last equation (2-58) to the first equation (2-59). We get an optimal specific load on the wing area

$$p_{opt} = \frac{\varsigma q}{g} \sqrt{\frac{C_{DW}}{\varepsilon}}. \tag{2-59}'$$

Substitute α from (2-58)' into the last equation (2-58)' and divide both sides by a vehicle mass m. We get

$$\frac{D}{m} = \left[C_{DW} \frac{1}{p} + \varepsilon \left(\frac{g}{\varsigma q}\right)^2 p \right] q. \tag{2-60}$$

Here: D/m is a specific drag (drag per a vehicle weight unit). Substitute (2-59)' into (2-60). We get the minimum drag of variable wing

$$\min\left(\frac{D}{m}\right) = 2 \frac{g}{\varsigma} \sqrt{\varepsilon C_{DW}}, \tag{2-60}'$$

where the member in the right side is wing drag per a lift of one vehicle weight unit. We discover the important fact: the OPTIMAL wing drag of a variable wing DOES NOT DEPEND on air speed. It depends ONLY on the geometry of a wing! It may look wrong, however consider the following example. Wing drag equals $D=mg/K$, where $K=C_L/C_D$ is the wing efficiency coefficient. The value D/m does not depend on speed.

If the air vehicle has body, the minimum drag is

$$\min\left(\frac{D}{m}\right) = 2 \frac{g}{\varsigma} \sqrt{\varepsilon C_{DW}} + C_{Db} \frac{q}{p_b}, \quad q = \frac{\rho V^2}{2}. \tag{2-61}$$

Full vehicle drag depends on a speed because the body drag depends from V.

Substitute (2-59)' to α into (2-58)'. We get the optimal attack angle

$$\alpha_{opt} = \sqrt{\frac{C_{DW}}{\varepsilon}}. \quad (2\text{-}62)$$

This is the angle of optimal efficiency, but C_{DW} is the wing drag coefficient ONLY when $\alpha = 0$ (not full vehicle as in the conventional aerodynamic). Coefficient of flight efficiency

$$k = \frac{g}{D/m} \quad or \quad k_{max} = \frac{g}{\min(D/m)}. \quad (2\text{-}63)$$

b) Hypersonic speed. Equation of wing lift force and wing air drag for hypersonic speed are the following:

$$Y = \varsigma \alpha q S = m\left(g - \frac{V^2}{R}\right), \quad or \quad \alpha = \frac{p(g - V^2/R)}{\varsigma q}, \quad \overline{D} = (C_{DW} + \varepsilon \alpha^2) q S, \quad q = \frac{\rho a V}{2}. \quad (2\text{-}64)$$

Substitute α from (2-64) into $\overline{D}$. We get

$$D = \left[C_{DW} + \varepsilon \left(\frac{m(g - V^2/R)}{\varsigma q S} \right)^2 \right] q S. \quad (2\text{-}64)'$$

Substitute the wing load $p = m/S$ to (2-64)'. We get

$$\frac{D}{m} = \left[C_{DW} \frac{1}{p} + \varepsilon \left(\frac{g - V^2/R}{\varsigma q} \right)^2 p \right] q. \quad (2\text{-}65)$$

Find the minimum of the air drag D for p. Take the derivatives and set them equal to zero. We get

$$p_{opt} = \frac{\varsigma q}{(g - V^2/R)} \sqrt{\frac{C_{DW}}{\varepsilon}}. \quad (2\text{-}66)$$

Substitute (2-66) in (2-65). We get a minimum wing drag

$$\min\left(\frac{D}{m}\right)_w = \frac{2}{\varsigma}\left(g - \frac{V^2}{R}\right)\sqrt{\varepsilon C_{DW}}.$$

Sum of the minimum vehicle drag plus a body drag is

$$\min\left(\frac{D}{m}\right) = \frac{2}{\varsigma}\left(g - \frac{V^2}{R}\right)\sqrt{\varepsilon C_{DW}} + C_{Db}\frac{q}{p_b}, \quad q = \frac{\rho a V}{2}, \quad p_b = \frac{m}{S_b}. \quad (2\text{-}67)$$

Substitute (2-66) to α into (2-64). We get the optimal attack angle of vehicle without body

$$\alpha_{opt} = \sqrt{C_{DW}/\varepsilon}. \quad (2\text{-}68)$$

Coefficient of the flight efficiency $k=Y/D$ is

$$k = \frac{g - V^2/R}{D/m}, \quad k_{max} = \frac{g - V^2/R}{\min(D/m)}.$$

For hypersonic speed the coefficients approximately equal

$$\varsigma = 4, \quad \varepsilon = 2, \quad C_{DW} = 4c^2, \quad C_{Db} = 2c_1^2, \quad C_L = \varsigma\alpha, \quad C_{Do} = C_{DW} + C_{Db}, \tag{2-69}$$

In numerical computation the angle θ can be found from (2-21) as $\theta = \Delta h / \Delta R_g$.

For the rocket engine or a gliding flight we find the following relation:

When S is optimum (variable), the partial derivatives from (2-67) equal

$$D_V' = -\frac{4V}{\varsigma R}\sqrt{\varepsilon C_{DW}} + C_{Db}\frac{\rho a}{2p_b}, \quad D_h' = -\frac{C_{Db}\rho a V}{2b_2 p_b}.$$

Substitute them in (2-33). We find the relationship between speed, altitude, and optimal wing load for a hypersonic vehicle with rocket engine and VARIABLE optimal wing:

$$\rho = \frac{8gp_b V\sqrt{\varepsilon C_{DW}}}{\varsigma a C_{Db} R(g + V^2/b_2)}, \quad h = 11000 + b_2 \ln\frac{a_2}{\rho}. \tag{2-70}$$

For $\varsigma=4$, $\varepsilon=2$ the equation (2-69)' has the form

$$\rho = \frac{2gp_b V\sqrt{2C_{DW}}}{C_{Db} aR(g + V^2/b_2)}, \quad h = 11000 + b_2 \ln\frac{a_2}{\rho}, \tag{2-70}'$$

Result of computation by (2-70)' for $\varsigma=4$, $\varepsilon=2$, $a_2=0.365$, $b_2=6997$ and different p_b are presented in figs. 2-6 (dash lines). As you see, the variable area wing saves a kinetic energy, because its curve is located over an invariable (fixed) wing. This is advantageous only at the orbital speed (7.9 km/sec) because no lift force is necessary.

Estimation of flight range.

Air and space vehicles without trust

Aircraft range can be found from Eq. (2-22)

$$R_a = \int_{V_1}^{V_2} \frac{mVdV}{T - D - mg\theta}, \quad V_1 > V_2 \quad \text{or} \quad R_a = \int_{V_2}^{V_1} \frac{VdV}{D/m + g\theta}, \quad \text{if} \quad T = 0., \tag{2-71}$$

Consider a missile having the *optimal variable wing* in a descent trajectory with thrust $T=0$.

a) Make a simplest estimation using equation of a kinetic energy from the classical mechanics. Separate the flight in two stages: hypersonic and subsonic. If we have the ratio of vehicle efficiency $k_1 = C_L/C_D$, $k_2 = C_L/C_D$, where k_1, k_2 ratio of flight efficiency for hypersonic and subsonic stages respectively, we find the following equations for a range in each region:

$$\frac{m}{2}\left(V_1^2 - V_2^2\right) = \frac{m(g - V^2/R)}{k_1}R_1, \quad R_1 = \frac{k_1\left(V_1^2 - V_2^2\right)}{2(g - V^2/R)}, \quad R_2 = k_2 h, \quad R_a = R_1 + R_2,$$

Or more exactly

$$d\left(\frac{mV^2}{2}\right) = \frac{m(g - V^2/R)}{k_1}dR_1, \quad R_1 = -\frac{k_1 R}{2}\ln\left(\frac{g - V_2^2/R}{g - V_1^2/R}\right), \qquad (2\text{-}72)$$

where R_1 is the hypersonic part of the range, R_2 is the subsonic part of the range, V_1 is an initial (maximum) vehicle hypersonic speed, V_2 is a final hypersonic speed, *and h* is an altitude in the initial stage of the subsonic part of trajectory.

b) To be more precise. Assume in (2-71) ρ=const (used average air density).

1. Hypersonic part of the trajectory. Substitute (2-67) to (2-71). We have

$$R_{1H} = \int_{V_1}^{V_2} \frac{VdV}{aV^2 + bV + c}, \quad \text{or} \quad R_{1H} = \int_{V_1}^{V_2} \frac{VdV}{X}, \quad \text{where} \quad X = aV^2 + bV + c,$$

$$a = \frac{2\sqrt{\varepsilon C_{DW}}}{\varsigma R}, \quad b = -C_{Db}\frac{\rho a}{2p_b}, \quad c = \frac{T}{m} - \frac{2g}{\varsigma} - g\theta, \quad \Delta = 4ac - b^2,$$

$$R_{1H} = \left[\frac{1}{2a}\ln X - \frac{b}{2a}\int\frac{dV}{X}\right]_{V_1}^{V_2}, \quad \int\frac{dV}{X} = \frac{2}{\sqrt{\Delta}}\arg\tan\frac{2aV + b}{\sqrt{\Delta}} \quad \text{for} \quad \Delta \geq 0, \qquad (2\text{-}73)$$

$$\int\frac{dV}{X} = -\frac{2}{\sqrt{-\Delta}}\arg\tanh\frac{2aV + b}{\sqrt{-\Delta}} = \frac{1}{\sqrt{-\Delta}}\ln\frac{2aV + b - \sqrt{-\Delta}}{2aV + b + \sqrt{-\Delta}} \quad \text{for} \quad \Delta \leq 0.$$

2. Subsonic part of the trajectory. Substitute (2-61) in (2-71). We get

$$R_{1S} = -\frac{1}{2C_2}\ln\left|\frac{C_1 - C_2 V_2^2}{C_1 - C_2 V_1^2}\right|, \qquad (2\text{-}74)$$

where the values C_1, C_2 are

$$C_1 = \frac{T}{m} - g\left(\frac{2\sqrt{\varepsilon C_{DW}}}{a\varsigma} + \theta\right), \quad C_2 = C_{Db}\frac{\rho}{2p_b}. \qquad (2\text{-}75)$$

The trajectory (without rocket part of trajectory) is

$$R_t = R_{1H} + R_{1S} \quad \text{or} \quad R_g = R_{1H} + R_{1S} + R_2. \qquad (2\text{-}76)$$

where $R_2 = k_2 h$ computed for altitude *h* at the end of a kinetic part of the subsonic trajectory.

3. Ballistic trajectory of a wingless missile without atmosphere drag is

$$h = \frac{gt^2}{2}, \quad t = \sqrt{\frac{2h}{g}}, \quad R_b = V_1 t = V_1\sqrt{\frac{2h}{g}}, \quad V_i = \sqrt{V_1^2 + V_y^2}, \quad V_y^2 = 2h(g - V^2/R), \qquad (2\text{-}77)$$

where h is the initial altitude, V_1 is the initial horizontal speed of the wingless missile at altitude h, V_y is an initial (shot) vertical speed at $h = 0$, V_i is the full initial (shot) speed at $h = 0$.

For hypersonic interval $5<V<7.5$ km/sec, we can use the more exact equation

$$R_b = V_1 \sqrt{\frac{2h}{(g - V_1^2 / R)}}, \qquad (2\text{-}78)$$

where $R=6378$ km is the radius of Earth. The full range of a ballistic rocket plus the range of a winged missile equals

$$R_f = R_b + R_a + R_g, \qquad (2\text{-}79)$$

where $R_g = kh$ is a vehicle gliding range from the final altitude h_2 (fig.5-1) with aerodynamic efficiency k.

The classical method of the optimal shot ballistic range for spherical Earth without atmosphere is

$$R_b = 2R\beta_{opt}, \quad \tan\beta_{opt} = \frac{v_A}{2\sqrt{1-v_A}}, \quad v_A = \frac{V_A^2}{V_c^2}, \qquad (2\text{-}80)$$

where β_{opt} is an optimal shot angle, V_A is a shot projectile speed, *and V_c is an orbital speed for a circle orbit at a given altitude.*

4. Cannon projectile. We divide the distance in three sub distances: *1) $1.2M < M$, 2) $0.9M < M < 1.2M$, 3) $0 < M < 0.9M$.* The range of the wing cannon projectile may be estimated by equation

$$R = \frac{k_1}{2g}(V_1^2 - V_2^2) + \frac{k_2}{2g}(V_2^2 - V_3^2) + \frac{k_3}{2g}(V_3^2 - V_0^2), \quad \text{where} \quad 0 < V_0 < V_3 < V_2 < V_1. \qquad (2\text{-}81)$$

where k_1, k_2, k_3 are average aerodynamic efficiencies for 1, 2, 3 sub distances respectively. Conventionally, these coefficients have values: subsonic k_3=8-15, near sonic k_2=2-3, supersonic and hypersonic k_1= 4-9. If $V > 600$ m/sec, the first member in (2-81) has the most value and we can use the more simple equation for range estimation:

$$R = \frac{k_1}{2g}V_1^2. \qquad (2\text{-}80)'$$

At the top of the trajectory, modern projectile can have an additional impulse from small rocket engines. Their weight equals 10-15% of the full mass projectile and increases the maximum range in 7-14 km. In this case we must substitute $V=V_1+dV$ in (2-80)', where dV is the additional impulse (150-270 m/sec).

Subsonic aircraft with trust. Horizontal flight

The optimal climb and descent of a subsonic aircraft with a constant mass and fixed wing is described by equations (2-46), (2-43). Any given point in a climb curve may be used for horizontal flight (with different efficiency). We consider in more detail the horizontal flight when the aircraft mass decreases because the fuel is spent. This consumption may reach 40% of the initial aircraft mass. The optimal horizontal flight range may be computed in the following way:

$$dR = V dt, \quad dt = \frac{dm}{c_s T} = \frac{g\, dm}{c_s D}, \quad dR = \frac{gV}{c_s D(m)} dm, \quad R = \frac{gV}{c_s} \int_{m_k}^{m} \frac{dm}{D(m)}, \qquad (2\text{-}82)$$

where m is fuel mass, c_s is fuel consumption, kg/sec/ kg trust.

a) For fixed wing, we have (from (2-40))

$$D = C_{Do}qS + \frac{\varepsilon}{qS}\left(\frac{g}{\varsigma}\right)^2 m^2, \quad \text{where} \quad C_{Do} = C_{Dw} + C_{Db}\left(\frac{S_b}{S}\right), \quad q = \frac{\rho V^2}{2}. \quad (2\text{-}83)$$

Substitute (2-83) in (2-82), we get

$$R = \frac{gV}{c_s\sqrt{C_1 C_2}}\arg\tan\frac{\sqrt{C_1/C_2}(m-m_k)}{1+(C_1/C_2)mm_k}, \quad \text{where} \quad C_1 = \frac{\varepsilon}{qS}\left(\frac{g}{\varsigma}\right)^2, \quad C_2 = C_{Do}qS, \quad (2\text{-}84)$$

b) For variable wing we have (from (2-61)

$$R = \frac{gV}{c_s C_1}\ln\frac{C_1 m - C_2}{C_1 m_k - C_2}, \quad \text{where} \quad C_1 = 2\frac{g}{\varsigma}\sqrt{\varepsilon C_{DW}}, \quad C_2 = C_{Db}qS_b, \quad \rho = \rho_0 e^{-h/b_1}, \quad (2\text{-}85)$$

Results of the computation are presented in fig. 2-7. Aircraft have the following parameters: C_{DW} = 0.02; C_{Db} =0.08; b_1 =9086; S =120 m²; m =100 tons, m_k =80 tons, c_s = 0.00019 kg/sec/kg trust; wing ratio λ =10.

As you see, the specific fuel consumption does not depend on speed and altitude, a good aircraft design reaches the maximum range only at one point, in one flight regime: when the aircraft flies at the maximum speed admissible by critical Mach number, at maximum altitude admissible by engine. The deviation from this point decreases the range in 5-10-15 percent or more. The variable wing increases efficiency of the other regime. That approximately decreases the losses by a half.

The coefficient of the flight efficiency may be computed by equation $k=g/(D/m)$, where values

$$\frac{D}{m} = C_{DW}\frac{q}{p} + \frac{\varepsilon p}{q}\left(\frac{g}{\varsigma}\right)^2 + C_{Db}\frac{q}{p_b}, \quad \left(\frac{D}{m}\right)_1 = 2\frac{g}{\varsigma}\sqrt{\varepsilon C_{DW}} + C_{Db}\frac{q}{p_b}, \quad (2\text{-}86)$$

for fixed and variable wings respectively. Result of computation is presented in fig. 2-8. The curve of the variable wing is the round curve of the fixed wing.

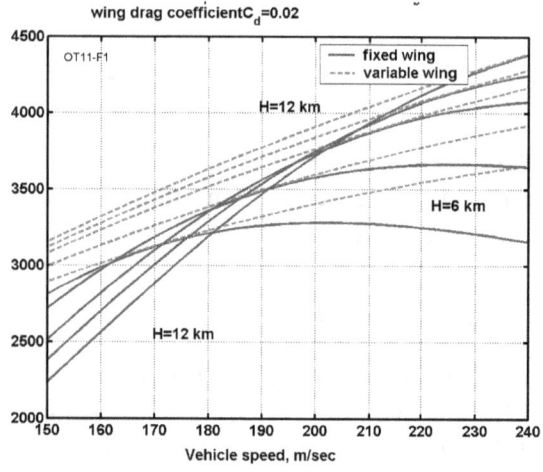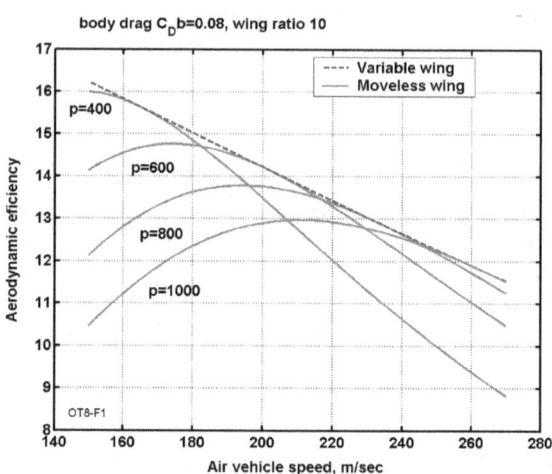

Fig. 2-7 (left). Aircraft range for altitude H = 6, 8, 10, 11, 12 km; maximum range R_m = 4361 km; relative fuel mass M_r = 0.2; body drag coefficient C_b = 0.08; wing drag coefficient C_d = 0.02.

Fig. 2-8 (right). Aerodynamic efficiency of non-variable and variable wings for wing load p = 400, 600, 800, 1000 kg/m², wing drag C_D = 0.02, body drag C_{Db} = 0.08, wing ratio 10.

3. Optimal engine control for constant flight pass angle

Let as to consider the equation (1-1)-(1-5) for constant trajectory angle θ = const. Substitute θ = constant, thrust $T=V_e\beta$, and a new independence variable $s = Vt$ (where s is the length of trajectory) into equation system (1-1)-(1-5). We get equations

$$\frac{dL}{ds} = \cos\theta ,$$

$$\frac{dh}{ds} = \sin\theta ,$$

$$\frac{dV}{ds} = \frac{V_e(h,V)\beta - \overline{D}(\alpha,V,h)}{mV} - \frac{g}{V}\sin\theta , \qquad (3\text{-}1)\text{-}(3\text{-}6)$$

$$\frac{dm}{ds} = -\frac{1}{V}\beta ,$$

$$Y(\alpha,V,h) - gm\cos\theta + \frac{mV^2}{R} + 2mV\omega_E \cos\varphi_E = 0 ,$$

$$0 \le \beta \le \beta_{max} .$$

The equation (3-5) is used for substitute α into equation (3-3) and for a change of air drag

$$\overline{D}(\alpha,V,h) = D(V,h). \qquad (3\text{-}7)$$

We received a nonlinear system with a linear fuel control β. It means this system can have a singular solution.

Solution

Consider the maximum range for vehicles described by system (3-1)-(3-6).
Let as to write Hamiltonian H

$$H = \cos\theta + \lambda_1 \sin\theta + \lambda_2 \left[\frac{V_e(h,V)\beta - D(V,h)}{mV} - \frac{g}{V}\sin\theta\right] - \lambda_3 \frac{1}{V}\beta , \qquad (3\text{-}8)$$

where $\lambda_1(s), \lambda_2(s), \lambda_3(s)$ are unknown multipliers. Application of conventional method gives

$$\dot{\lambda}_2 = -\frac{\partial H}{\partial V} = -\lambda_2\left[\left(-\frac{1}{V^2}\right)\left(\frac{V_e\beta - D(V,h)}{m}\right) - g\sin\theta\right) - \frac{D'_V}{mV}\right] - \lambda_3 \frac{1}{V^2}\beta ,$$

$$\dot{\lambda}_3 = -\frac{\partial H}{\partial m} = \lambda_2 \frac{V_e\beta - D}{m^2 V} , \qquad (3\text{-}9)\text{-}(3\text{-}11)$$

$$\beta = \max_{\beta} H = \beta_{max} \text{sign}[\lambda_2 V_e - \lambda_3 m] .$$

Where D'_V is the first partial derivates D per V.

The last equation shows that a fuel control β can have only two values, $\pm\beta_{max}$. We consider the singular case when

$$A = \lambda_2 V_e - \lambda_3 m \equiv 0 . \qquad (3\text{-}12)$$

This equation has two unknown variable, λ_2 and λ_3, and does not contain information about fuel control β. The first two equations (3-1)-(3-2) do not depend on variable and can be integrated

$$L = s \cos\theta, \quad (3\text{-}13)$$

$$h = s \sin\theta. \quad (3\text{-}14)$$

According with book [2] let as to differentiate the equation (3-12) to the independent variable s. After substitution the equations (3-3)-(3-5), (3-7), (3-9), (3-10), (3-12), (3-14) we get the relation for $\lambda_2 \neq 0$, $\lambda_3 \neq 0$:

$$\boxed{\dot{A} = VD - mVD'_m + V_e(-D - mg\sin\theta + VD'_V) - VV'_{e,V}(D - mg\sin\theta) + mV^2 V'_{e,s} = 0.} \quad (3\text{-}15)$$

This equation also does not contain β, however it contains an important relation between variables m, h and V, on an optimal trajectory. This is 3-dimentional surface. If we know

$$D = D(h, V), \quad (3\text{-}16)$$

$$V_e = V_e(h, V), \quad (3\text{-}17)$$

mass of our apparatus m, and it's altitude h, we can find the optimal flight speed. It means we can know the needed thrust and the fuel consumption for every points m, h, V (Fig.3-1).

If we want to have an equation for a fuel control β we continue to differentiate the equation (3-15) to the independence variable s and substitute equations (3-1)-(3-14). We calculate the relation for β if $\lambda_2 \neq 0$, $\lambda_3 \neq 0$, V_e=const, then

$$\beta = \frac{\dot{A}_V(D + mg\sin\theta) - mV\dot{A}_s}{V_e \dot{A}_V - m\dot{A}_m}, \quad (3\text{-}18)$$

where

$$\dot{A}_V = \frac{\partial}{\partial V}\left(\frac{dA}{ds}\right), \quad \dot{A}_s = \frac{\partial}{\partial s}\left(\frac{dA}{ds}\right). \quad (3\text{-}19)$$

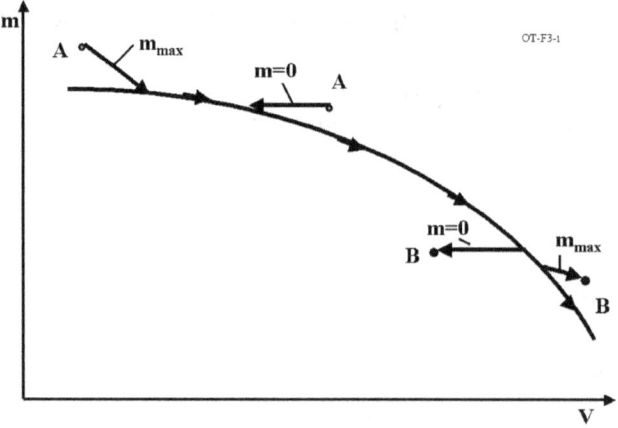

Fig.3-1. Optimal fuel consumption of flight vehicles.

The necessary condition of the optimal trajectory as it is shown in [2]-[8] (see for example, [4], Eq. (4.2)) is

$$-(-1)^k \frac{\partial}{\partial \theta}\left[\frac{d^{2k}}{ds^{2k}}\left(\frac{\partial H}{\partial s}\right)\right] \geq 0. \tag{3-20}$$

where $k = 1$.

If the flight is horizontal ($\theta = 0$), the expression (3-18) is very simply

$$\beta = \frac{D}{V_e}. \tag{3-21}$$

It means, the trust equals the drag. This fact is well known in aerodynamic science.

For getting the specific equations for different forms of drag and thrust, we must take formulas (or graphs) for a subsonic, transonic, supersonic and hypersonic speed for a thrust and substitute them in the equation (3-15), (3-18).

4. Simultaneously optimization of the path angle and fuel consumption

Consider the case when the path angle and the fuel consumption are simultaneously optimized. In this case the general equations (1-1)-(1-5) have a form:

$$\frac{dL}{ds} = 1,$$

$$\frac{dh}{ds} = \theta,$$

$$\frac{dV}{ds} = \frac{V_e(h,V)\beta - D(m,V,h)}{mV} - \frac{g}{V}\theta,$$

$$\frac{dm}{ds} = -\frac{1}{V}\beta, \tag{4-1)-(4-5)}$$

$$Y(\alpha, V.h) = mg + \frac{mV^2}{R} + 2mV\omega_E \cos\varphi_E.$$

Let us to write the Hamiltonian

$$H = 1 + \lambda_1\theta + \lambda_2\left(\frac{V_e(h,V)\beta - D(m,V,h)}{mV} - \frac{g}{V}\theta\right) - \lambda_3\frac{1}{V}\beta. \tag{4-6}$$

The necessary conditions of optima give

$$A = \frac{\partial H}{\partial \theta} = V\lambda_1 - g\lambda_2 = 0,$$

$$B = \frac{\partial H}{\partial \beta} = V_e\lambda_2 - m\lambda_3 = 0, \tag{4-7)-(4-8)}$$

The lamda equations are

$$\dot{\lambda_1} = -\frac{\partial H}{\partial h} = -\lambda_2 \frac{V'_{e,h}\beta - D'_h}{mV},$$

$$\dot{\lambda_2} = -\frac{\partial H}{\partial V} = -\lambda_2\left[\frac{(V'_{e,V}\beta - D'_V)V - (V_e\beta - D)}{mV^2} + \frac{g}{V^2}\theta\right] - \lambda_3\frac{1}{V^2}\beta, \quad (4\text{-}9)\text{-}(4\text{-}11)$$

$$\dot{\lambda_3} = -\frac{\partial H}{\partial m} = \lambda_2 \frac{V_e\beta - D + mD'_m}{m^2V}.$$

Let us to difference A (4-7). From $dA/ds=0$, we find the optimal fuel consumption

$$\boxed{\beta = \frac{gVD'_V - V^2 D'_h}{g(V_e + V'_{e,V}V) - V'_{e,h}V^2}.} \quad (4\text{-}12)$$

Let us difference B (4-8). From $dB/ds=0$ we find the optimal path angle

$$\boxed{\theta = \frac{V'_{e,V}D - V_e D'_v - V_e D/V - D + mD'_m}{m(g + V'_{e,h}V - V'_{e,V}g)}.} \quad (4\text{-}13)$$

We used the conventional marks for the partial derivatives in (4-9)-(4-13) as in part 2 and 3 (see for example (2-47)).

If we know from analytical formula or graphical functions Ve, D, Y we can find the optimal trajectory of the air vehicle.

In the general case, this trajectory includes four parts:

1. Moving between limitations θ and β.
2. Moving between one limitation θ or β and one optimal control β or θ.
3. Moving simultaneously with both optimal controls θ and β.
4. Moving on a given point along one limitation and/or both limitations.

5. Application to aircraft, rocket missiles, and cannon projectiles

A) Application to rocket vehicles and missile.

Let us to apply the previous results to the typical current middle and long distance rockets with warhead. We will show: if warhead has wings and uses the optimal trajectory, the range of warhead (or useful load) is increased dramatically in most cases. We will compute the following optimal trajectories: the rocket launched warhead beyond altitude (20 – 60 km) and speed (1 – 7.5 km/sec). The point B is located on the curve (2-54) for a fixed wing and it is located on curve (2-69)' for a variable wing (fig.5-1). Further, the winged warhead flights (descent) along optimal trajectory BD (fig.5-1) accorded the equations (2-54)(fixed wing) or the equations (2-69)'(variable wing) respectively. When speed is decreased a small amount (for example, 1 km/sec) (point D in fig. 5-1), the wing warhead glides (distance DE in fig.5-1).

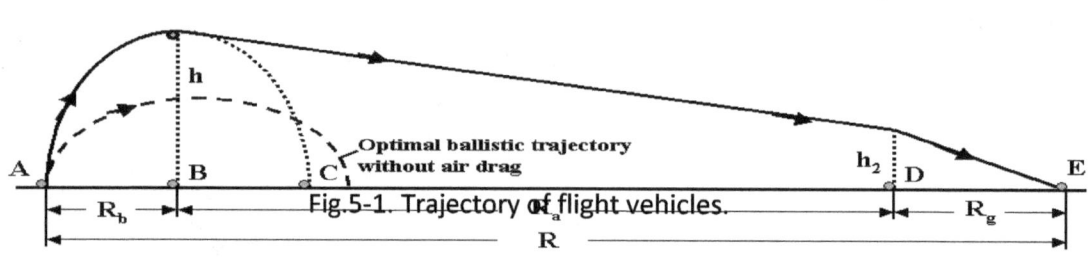

Fig.5-1. Trajectory of flight vehicles.

The following equations are used for computation:

1. *The optimal trajectory for space vehicle with FIXED wing.*
   a) Eq. (2-54) is used for computing $h=h(V)$ of the warhead optimal trajectory of the non- variable fixed wing in the speed interval $1<V<7.5$ km/sec. Result is presented in fig.2-6.
   b) Eq. (2-50) computes the magnitude (D/m).
   c) The equation (2-71) in form

$$\Delta R_a = \frac{V\Delta V}{(D/m)+g\theta}, \quad R_a = \Sigma \Delta R_a, \quad \theta = -\frac{\Delta h}{\Delta R_a}, \quad k = \frac{g-V_0^2/R}{D/m}, \quad R_g = h_0 k, \quad (5\text{-}1)$$

   is used for computation in the intervals R_a, R_g fig.5-1. Here R_g is the range of a gliding vehicle.
   d) The equation (2-71) is used for computation R_b in the launch interval AB fig. 5-1.
   e) The full range, R, of warhead with a fixed wing and the full ballistic warhead range, R_w, are

$$R = R_b + R_a + R_g, \quad R_w = 2R_b. \qquad (5\text{-}2)$$

The equation (2-80) is used for computation the optimal shot BALLISTIC trajectory without air drag (vehicle WITHOUT wing). The range of this trajectory, as it is known, may be significantly more then range in the atmosphere.

Result is presented in fig. 5-2. Computation of a relative range (for different p_b) by formulas

$$R_r = \frac{R_f}{R_b} \qquad (5\text{-}3)$$

is presented in fig. 5-3. The optimal range of the winged vehicle is approximately 4.5 times that of the ideal ballistic rocket computed without air drag. In the atmosphere this difference will be significantly more.

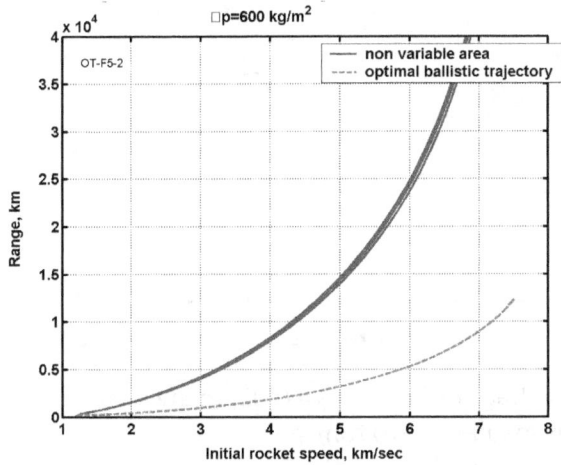
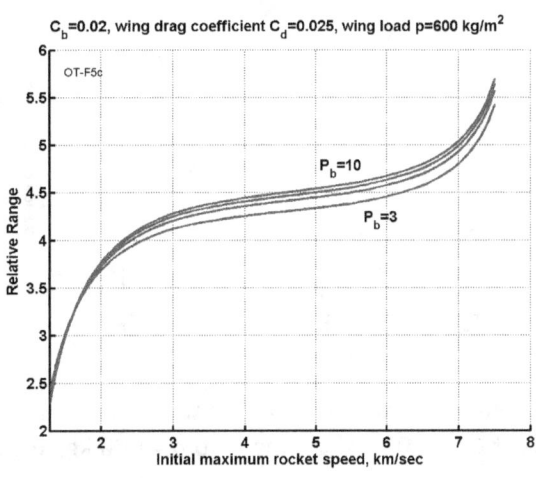

Fig.5-2 (left). Range of NON-VARIABLE wing vehicle for body drag coefficient C_b = 0.02, wing drag coefficient C_d = 0.025, wing load p = 600 kg/m².

Fig.5-3 (right). Relative range of NON-variable wing vehicle for the body drag coefficient C_b = 0.02, the wing drag coefficient C_d = 0.025, the wing load p = 600 kg/m², the body load P_b=3 –10 ton/m².

2. Rockets, missiles and space vehicles with VARIABLE wings.

The computation is same. For computing ρ, h, D/m we are used the equations (2-69)', (2-67) respectively. The results for different body loads are presented in fig.2-6. The optimal trajectories of vehicles with the variable wing areas have a lesser slope. It means, the vehicle loses less energy when it moves. It is located over the optimal trajectory of the vehicle with fixed wings. It means it needs a lot of time (10-20) and more wing area then the fixed wing space vehicle (fig.5-4). The computation of the optimal variable wing area is presented in fig. 5-5. The relative range (Eq. (5-3)) is presented in fig. 5-6.

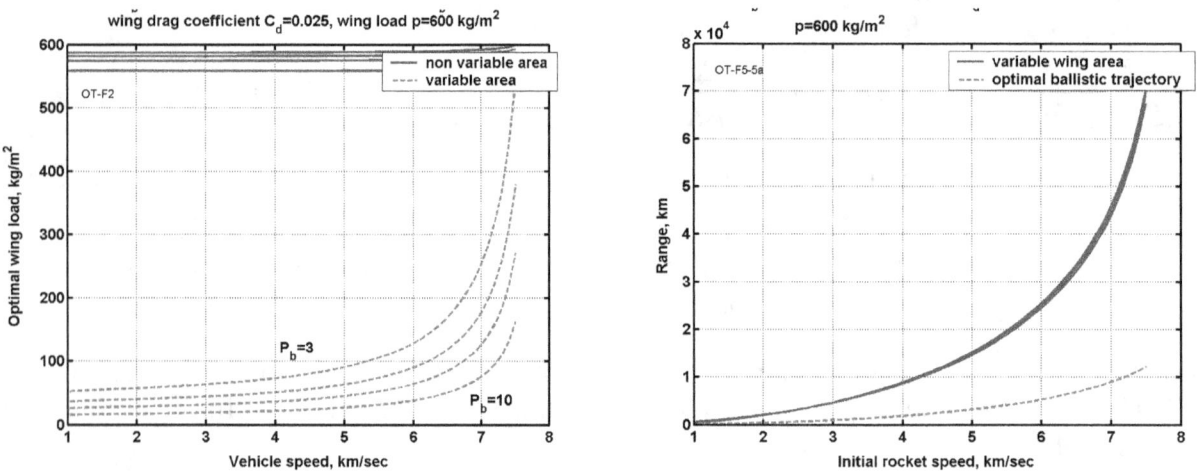

Fig.5-4 (left). Optimal wing load versus speed for specific body load P_b = 3, 5, 7, 10 ton/m², body drag coefficient C_b = 0.02, wing drag coefficient C_d = 0.025, wing load p = 600 kg/m².

Fig.5-5 (right). Range of VARIABLE wing vehicle for the body drag coefficient C_b = 0.02, the wing drag coefficient C_d = 0.025, the wing load p = 600 kg/m².

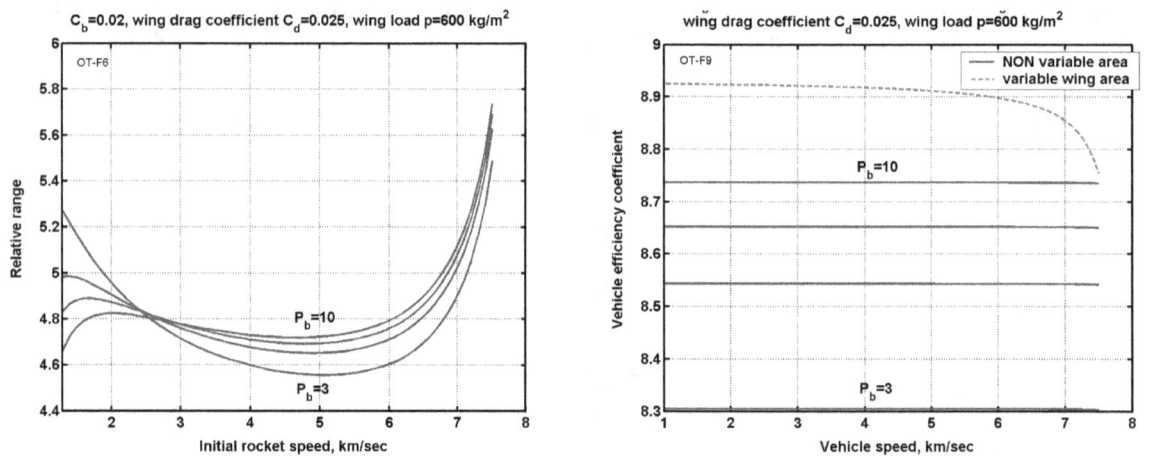

Fig.5-6 (left). Relative range of variable wing vehicle for the body drag coefficient C_b = 0.02, the wing drag coefficient C_d = 0.025, the wing load p = 600 kg/m², the body load P_b=3 –10 ton/m².

Fig.5-7 (right). Vehicle efficiency coefficient versus speed for specific body load P_b = 3, 5, 7, 10 ton/m², body drag coefficient C_b = 0.02, wing drag coefficient C_d = 0.025, wing load p = 600 kg/m².

The aerodynamic efficiency of vehicles with fixed (for different p_b bodies) and optimal variable wings computed by equations (5-1), (2-63) respectively are presented in fig. 5-7. The difference between the vehicle with fixed and variable wings reaches 0.2÷0.6 . The slope of the trajectory to horizontal is small (fig.5-8).

The range of the fixed wing vehicle computed by equation (5-1) is presented in fig. 5-2. The range of the variable wing vehicle computed by equation (5-2) is presented in fig. 5-5. The curve is practically same (see figs. 5-2, 5-5).

3. *Increasing of a rocket payload for same range.* If we do not need to increase the range, the winged vehicle can be used to increase payload, or save a rocket fuel. We can change the mass of fuel or payload. The additional payload my be estimated by equation

$$\mu = 1 - e^{-\frac{\Delta V}{V_e}}, \qquad (5-4)$$

where $\mu = m/m_b$ is a relative mass (the ratio of a rocket mass with wing vehicle to the ballistic rocket), $\Delta V = V_b - V$ is difference between the optimal ballistic rocket speed (Eq.(2-80)) and the rocket with winged vehicle (Eq.(5-2)) for given range (see fig.5-2). Result of computation is presented in fig. 5-9. The mass of the rocket with winged vehicle may be only 20 – 35% from the optimal ballistic rocket flown without air drag.

Conclusion: The winged air-space vehicle increases the range by a minimum of 4.5-5 times compared to a shot optimal ballistic space vehicle. The variable wing improves the aerodynamic efficiency 3-10% and also improves the range. The optimal variable wing requires a large wing area. If you do not need to increase the range, you may instead, increase payload.

B) Application to cannon wing projectiles.

The typical current cannon properties are shown in table 1 below.

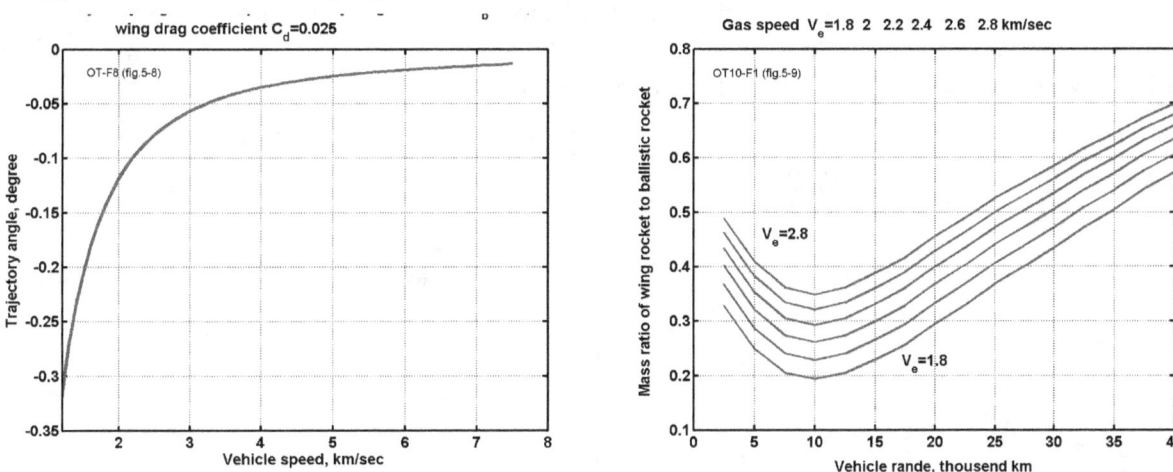

Fig.5-8 (left). Trajectory angle versus speed for body drag coefficient C_b =0.02, wing drag coefficient C_d = 0.025.

Fig.5-9 (right). Mass ratio of wing rocket to ballistic rocket for specific engine run-out gas speed V_e = 1.8, 2, 2.2, 2.4, 2.6 and 2.8 km/sec.

Table 1.

Name	caliber *mm*	Nozzle speed *m/sec*	Mass projectile *kg*	Range *km*	RAP *km*
M107	175	509-912	67	15-33	

SD-203	203	960	110	37.5	
2S19	155	810	43.6	24.7	
2S1	122	690-740	21.6	-	
S-23	180	-	-	30.4	43.8
2A36	152	-	-	17.1	24
D-20	152	600-670	43.5-48.8	20	

Issue: Jame's

The computation by Equation (2-80)' for different k and RAP with $dV=270$ m/sec are presented in figs. 5-10, 5-11.

Conclusion. As you see (figs.5-10, 5-11), the winged projectile increase range 3 – 9 times (from 35 up 360 km, $k = 9$). The projectile with RAP increase range 5-14 (from 40 up 620 km, $k=9$) times. The winged shells have another important advantage: they do not need to rotate. We can use a barrel with a smooth internal channel. This allows for the increase of projectile nozzle speed up 2 km/sec and the shell range up to 1000 km ($k=5$).

C) Application to current aircraft.

We used the equations (2-84), (2-85) for computation of the typical passenger airplane (Fig.5-12, thru 5-14, and 2-7). When all values are divided by maximum range R_m=4381 km (for a fuel mass 20% from a vehicle mass) at speed V=240 m/sec, altitude H=12 km. The speed is limited by the critical Mach number ($V<M$=0.82), the altitude is limited by the admissible engine trust, when engine stability is such that it works in cruiser regime. Fig. 5-12 shows the typical aircraft long-range trajectory.

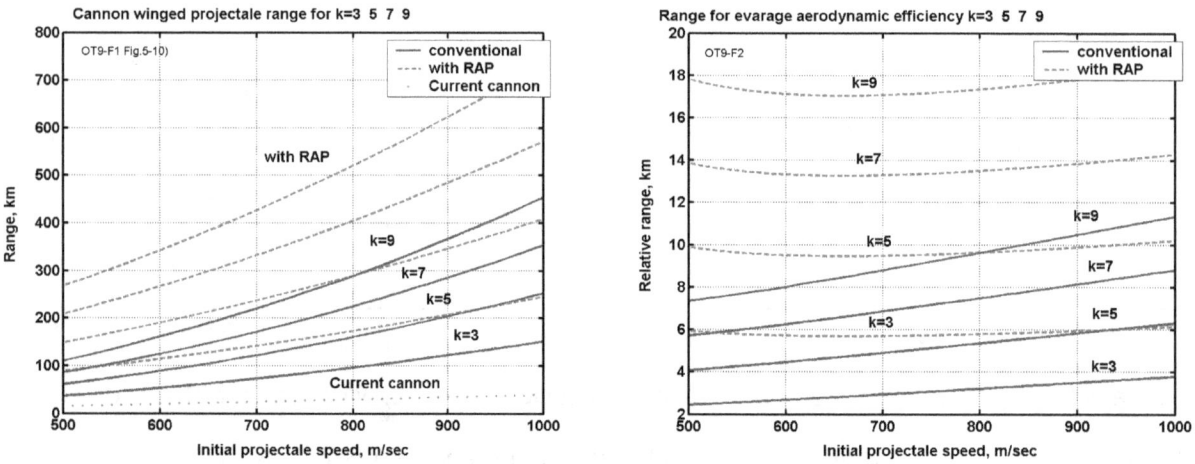

Fig.5-10 (left). Cannon winged projectile range for average aerodynamic efficiency k = 3, 5, 7, 9.
Fig.5-11 (right). Cannon winged projectile relative range for average aerodynamic efficiency k = 3, 5, 7, 9.

Conclusion: The best flight regime for a given air vehicle (closed to Boeing 737) is altitude H=12 km, speed V=240 m/sec, specific fuel consumption Cs=0.00019 kg fuel/sec/kg trust. The deviation from this flight regime significantly decreases the maximum range (up to 10-50%). The vehicle with a variable wing area losses 50% less then vehicle with a fixed wing.

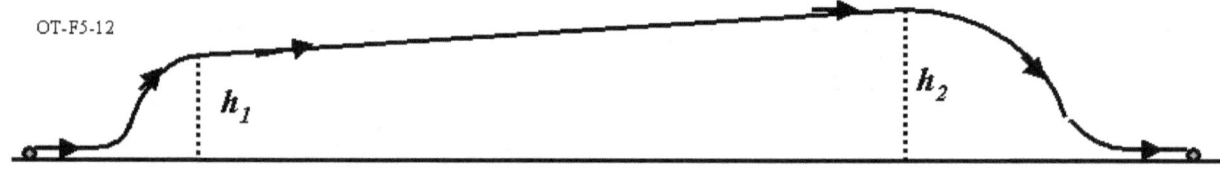

Fig.5-12. Optimal trajectory of aircraft.

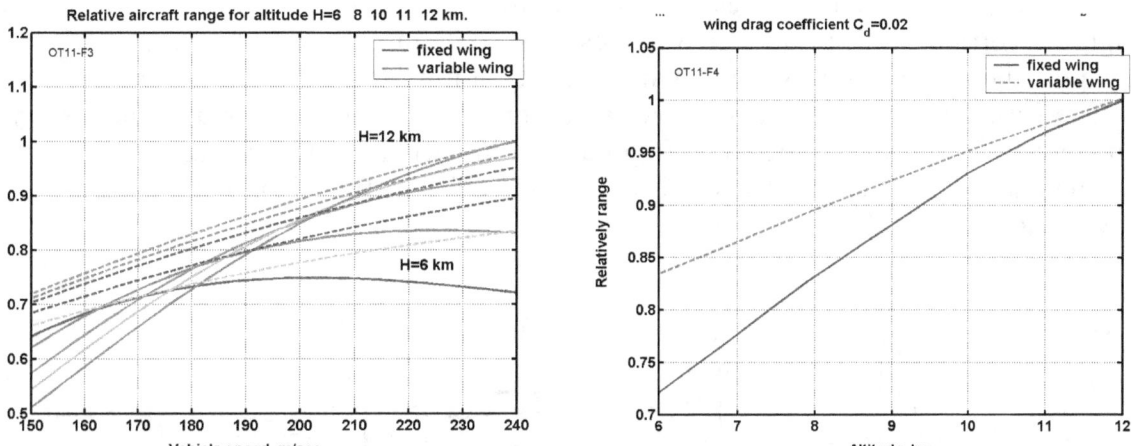

Fig.5-13 (left). Relative aircraft range for altitude H = 6, 8, 10, 11 and 12 km, maximum range R_m = 4381 km, relative fuel mass M_r = 0.2, body drag coefficient C_b = 0.08, wing drag coefficient C_d = 0.02.

Fig.5-14 (right). Relative aircraft range for speed V = 240 m/sec, maximum range R_m = 4381 km, relative fuel mass M_r = 0.2, body drag coefficient C_b = 0.08, wing drag coefficient C_d = 0.02.

6. General discussion and Conclusion

a.) The current space missiles were designed 30-40 years ago. In the past we did hot have navigation satellites that allowed one to locate a missile (warhead) as close as one meter. Missile designers used inertial navigation systems for ballistic trajectories only. At the present time, we have a satellite navigation system and cheap devices, which allow locating of aircraft, sea ships, cars, any vehicle, and people. If we exchange the conventional warhead by a warhead with a simple fixed WING, having a control and navigation system, we can increase the range of our old rockets 4.5 – 5 times (fig.5-3) or significantly increase the useful warhead weight (fig.5-9). We also notably improve the precision of our aiming.

b.) Current artillery projectiles for big guns and cannons were created many years ago. The designers assumed that the observer could see an aim point and correct the artillery. Now we have the satellite navigation system which allows one to get exact coordinates of targets and we have cheap and light navigation and control devises which can be placed in the cannon projectiles. If we replace our cannon ballistic projectiles by a projectile with a fixed wing and control, navigation system, we increase its range 3-9 times (from 35 km up 360 km, see fig. 5-10, 5-11). We can use a smooth barrel to increase nozzle shell speed up to 2000 m/sec and range up to 1000 km. These systems can guide the WINGED projectiles and significantly improving their aiming. We can reach this result because we use all of the KINETIC energy of the projectile. The conventional projectile cannot keep itself in the atmosphere and drops with a very high speed. Most of its kinetic energy is uselessly spent. In our case 70-85% of the projectile's kinetic energy is spent for support of the moving projectile. That way the projectile range increases 3-9 times or more.

c.) All aircraft are designed for only one optimal flight regime (speed, altitude, and fuel consumption). Any deviation from this regime decreases the aircraft range. For aircraft closed to the Boeing 747 this regime is: altitude H=12 km, speed V=240 m/sec, specific fuel consumption Cs =0.00019 kgf/sec/kg trust. If the speed is decreased from 240 m/sec to 200 m/sec, the range decreases 15% (fig.5-13). The application of the variable wing area decreases this loss from 15% to 10%. If the aircraft decreases the altitude from 12 km to 9 km, it loses 12% of its maximum range (fig.5-14). If it has a variable wing area, the air vehicle loses only 7.5% of maximum range. The civil air vehicles must deviate from the optimal condition by weather or a given flight air corridor. The military air vehicles must sometimes have a very large deviation from the optimal condition (for example, when they fly at low altitude out of the enemy radar system). The variable wing area may be very useful for them because it decreases the loss by approximately 50%, improves supersonic flight and taking off and landing lengths.

The author offers some fixed and variable wings for air vehicles (fig.5-15). Variants *a, b, c, f* for missile and warhead, variants *d, e* for shells.

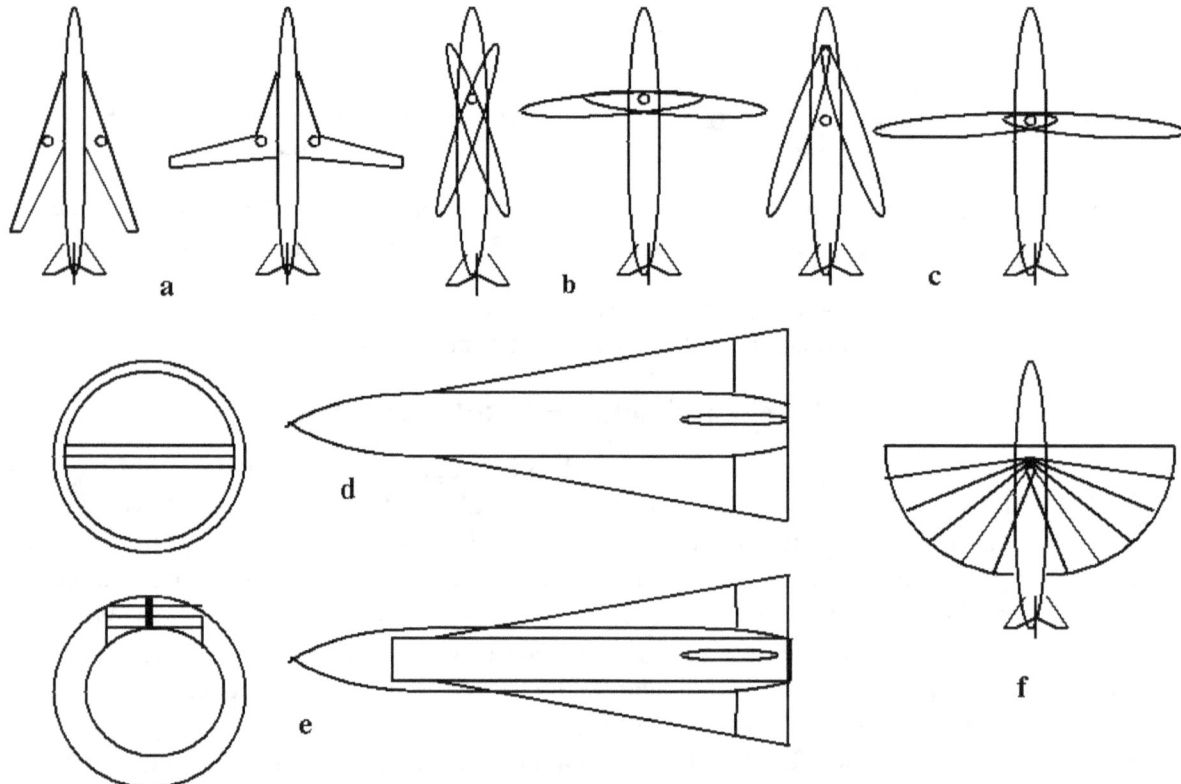

Fig.5-15. Possible variants of variable wing: a, b, c and f, *for aircraft;* d and e *for gun projectiles.*

References

1. A. Miele, General Variable Theory of the Flight Paths of Rocket-Powered Aircraft, Missile and Satellite Carries, *Astronautica acta*, Vol.4, pp.256-288, 1958.
2. A.A. Bolonkin, "Special Extrema in Optimal Control Problems", Akademiya Nauk, Izvestia, *Theknicheskaya Kibernetika*, No.2, March-April, 1969, pp.187-198. See also English translation in Eng. Cybernetics, No.2, March-April 1969, pp.170-183.
3. A.A. Bolonkin, New methods of optimization and their application, Moscow, Technical University named Bauman, 1972, pgs.220 (in Russian).
4. A.A. Bolonkin, "A New Approach to Finding a Global Optimum", "*New Americans Collected Scientific Reports*", Vol.1, 1991. The Bnai Zion Scientists Division, New York.
5. A. A. Bolonkin, Optimization of Trajectory of Multistage Flight Vehicles, in collection "*Researches of Flight Dynamics*", Masinobilding, Moscow 1965, pp.20-78 (in Russian).

6. A.A. Bolonkin, N.S. Khot, Method for Finding a Global Minimum, AIAA/NASA/USAF/SSMO Symposium on Multidisciplinary Analysis and Optimization, Panama City, Florida, USA. Sept.7-9, 1994.
7. A. A. Bolonkin, Solution Methods for Boundary-Value Problems of Optimal Control Theory. 1973. Consultants Bureau, a division of Plenum Publishing Corporation, NY. Translation from Prikladnaya Mekhanika, Vol.7, No 6, 1971, p.639-650 (English).
8. A. A. Bolonkin, Solution general linear optimal problem with one control. Journal *"Pricladnaya Mechanica"*, v.4, #4, 1968, pgs. 111-122, Moscow (in Russian).
9. A. A. Bolonkin, N. Khot, Optimal Structural Control Design, IAF-94-1.4.206, 45th Congress of the International Astronautical Federation, World Space Congress-1994. October 9-14,1994/Jerusalem, Israel.
10. A. A. Bolonkin, N. S. Khot, Optimal Bounded Control Design for Vibration Suppression, *Acta Astronautica*, Vol.38, No.10, pp.803-813, 1996.
11. A. A. Bolonkin, R. Sierakowski, Design of Optimal Regulators, 2nd AIAA "Unmanned Unlimited" Systems, Technologies, and Operations – Aerospace, Land, and sea Conference and Workshop & Exhibit, San Diego, California, 15-18 Sep 2003, AIAA-2003-6638.

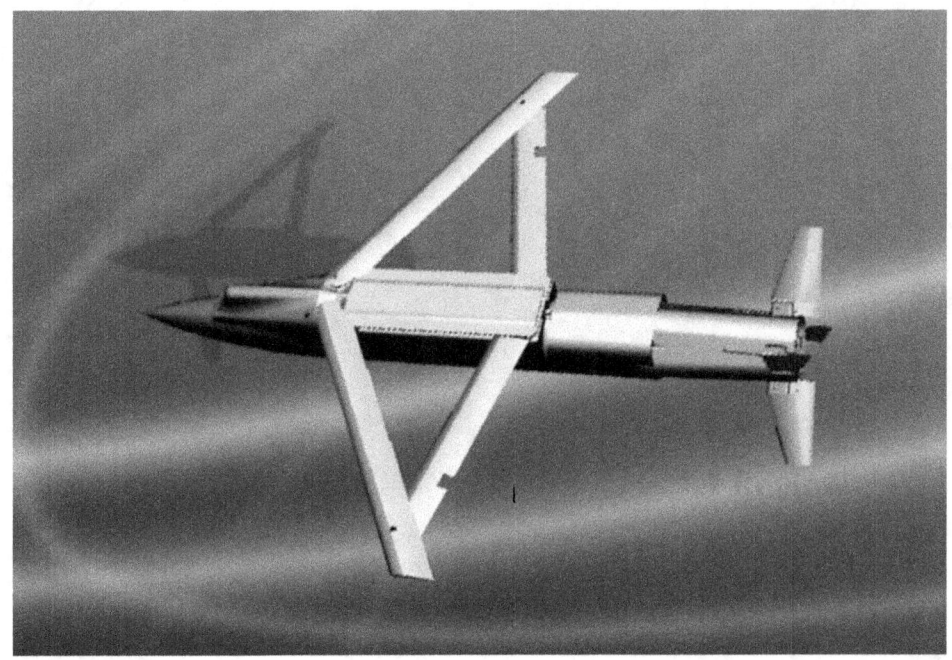

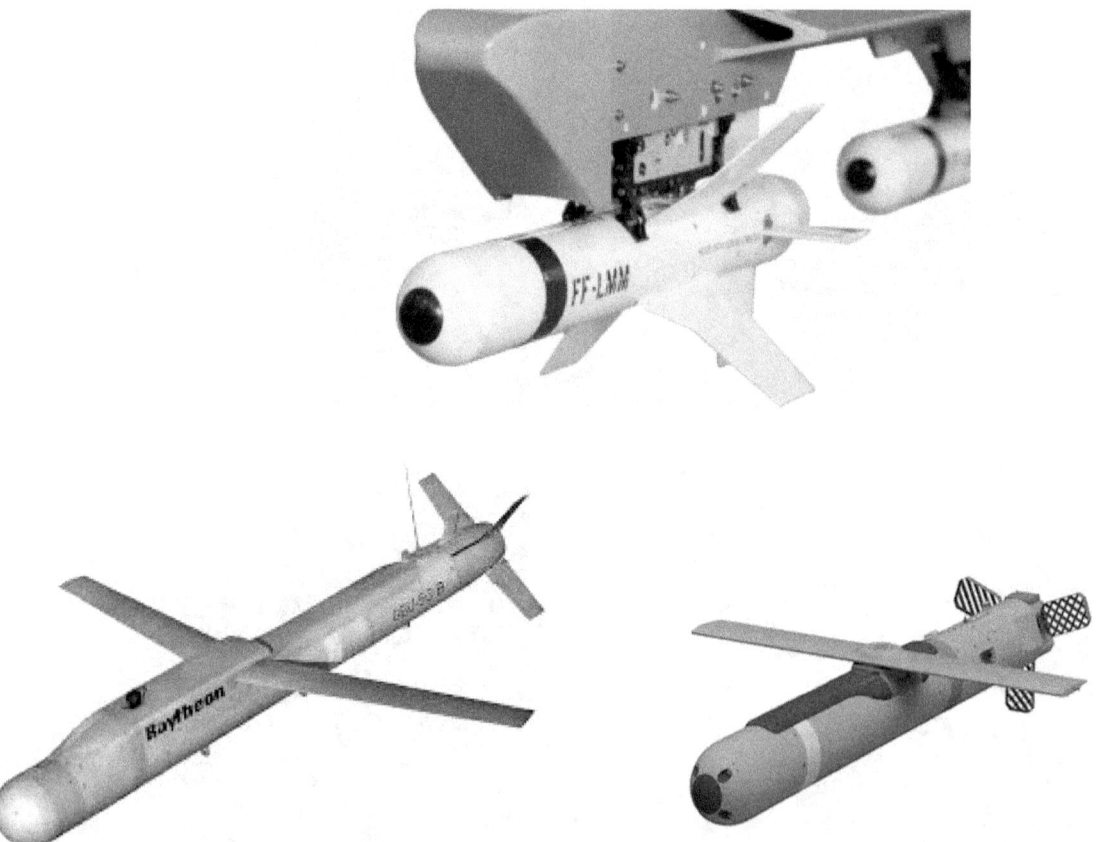

Chapter 14
Long Distance Artillery
Abstract.

This picks up on the author's early work of increasing range of the shells and bullets 2 – 5 times by including in its design small wings. The shell/bullet specially formed wings support the projectile in the air, does not allow it to fall in earth's surface as the kinetic energy the projectile is not spent fighting the forces of gravity and air resistance. This is an important innovation as it can be used in conventional rifles and gun with rifled barrel and rotary shell/bullet. The second idea is radical change of trajectory. The projectile reaches high altitude and glides from height using wings with subsonic speed and a good ratio lift/drag. Author developed theory of these projectile and computed some projects which show high efficiency of these innovations. This can be immediately integrated into the arms industry and army because it does not require new weapons (rifles, guns), but is a modification only of the bullets and shells.

Word keys: Wing projectile, wing shell, long distance shell, long distance bullet.

1. Introduction.

The idea of a wing artillery shell was first published in 1972 [1] – [2] with the full theory published in [3]-[7].

Guided Artillery Shell. An M982 Excalibur precision-guided artillery shell was developed on 50/50 basis by United States-based Raytheon Missile Systems (guidance system) and the Swedish BAE Systems Bofors (body, base, ballistics and payload) caliber is 155 mm. The "smart" round has a range of approximately 40 kilometers (25 mi) to 57 kilometers (35 mi) depending on configuration, with a circular error probable (CEP) of around 20 meters (66 ft.). The extended range is achieved through the use of folding glide fins, which allow the projectile to glide from the top of a ballistic arc towards the target. The accuracy is achieved through the use of a GPS guidance system. Typical (unguided) 155 mm shells have a CEP of 200 meters (660 ft.) to 300 meters (980 ft.) at moderate ranges.

The munitions was developed with $55.1 US million in financial assistance from Sweden, which expected to receive service rounds in 2010. As of 2008 unit cost was $85,000US, potentially dropping to $50,000US in full-scale production. The weapon can make first round strikes on targets up to 20 kilometers (12 mi) away.

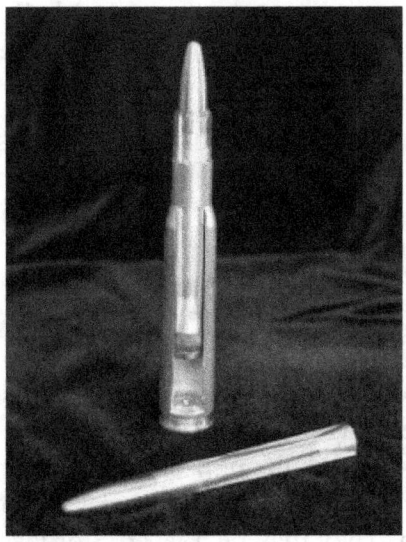 

Fig.1. (left) M982 Excalibur. A GPS guided artillery shell. (right) Sandia's bullet for special non riffled rifles and gun. Length of bullet is 10.2 cm, caliber about 12 mm.

Excalibur is used to minimize collateral damage, for targets beyond the range of standard munitions, for precise firing within 150 meters (490 ft.) of friendly troops, or when firing in a straight line from the launching cannon is limited by terrain.

The US Army rates Excalibur as one of the *Top 10 Army Greatest Inventions of the Year Award for 2007*. Initial combat experience with Excalibur in Iraq in the summer of 2007 was so successful, with 92% of rounds falling within 4 meters (13 ft) of the target that the US Army planned to increase production to 150 rounds per month from the previous 18 rounds per month.

Guided Bullet. Sandia National Laboratories engineers offered a dart-like, self-guided bullet for small-caliber, smooth-bore firearms that could hit laser-designated targets at distances of about mile (about 1600 meters)(2012).

Sandia's design needs a special gun. It uses the four-inch-long bullet (10.2 cm; caliber about 12 mm), smoothbore non rifled rifles and guns. One includes an optical sensor in the nose to detect a laser beam on a target. The sensor sends information to guidance and control electronics that use an algorithm in an eight-bit central processing unit to command electromagnetic actuators.

These actuators steer tiny fins that guide the bullet to the target. Most bullets are shot from rifles, which have grooves, or rifling, that cause them to spin so they fly straight, like a long football pass; to enable a bullet to turn in flight toward a target and to simplify the design, the spin had to go.

The bullet flies straight due to its aerodynamically stable design, which consists of a center of gravity that sits forward in the projectile and tiny fins that enable it to fly without spin.

Methods of targeting. There are some methods for navigation and targeting projectiles: GPS, laser beam, TV. Every method has advantages and disadvantages.

2. Description and Innovations

It is well-known that all bodies fall to the Earth. The force of gravity is so great that even a bullet/shell with enormous kinetic energy over a long distance will inevitably fall to the ground. To overcome the force of gravity, the author proposes to change the shape of the bullet/shell so that one has the lift force and remains in flight as long as there remains sufficient kinetic energy. This is not easy problem because all rifles and guns have rifling; bullet/shell rotates in flight (for stability) and no rotated form can produce a good lift force while in rotation. Author proposes a solution. The computations show the new form increases the bullet/shell range by 2 – 5 times! (last number for shell). A critical advantage of the new method is that it does not require new rifles, gun and cannon. The innovation is ONLY

the new form of bullet/shell and a possible new long-distance calibrated gun sight. This is by far the simplest and cheapest method for increasing range of the current weapon in 2 – 5 times. This new method needs financing to perfect the theory by computation and testing so that it can be used by the army in approximately 4 – 6 months.

The suggested bullet/shell is shown in fig.2. One has some modifications. The simplest variant is shown in fig. 2-(1). That has two small wings (2, 3) (forward and back).

The second version has reduced diameter caliber and discarding sabots (Fig.2 –(2)). That has small aerodynamic drag and longer range.

The third version (fig. 2-(3)) has the pull-out mobile variable sweep wings.
If we need in great accuracy, the projectile must have an optical sensor or/and a navigation system and guidance and control system. They may be TV, GPS or laser. Every system has their advantages and disadvantages. Author offers two new systems for shells. One system is the pattern recognizing of

target, the other system shows the result of fire.

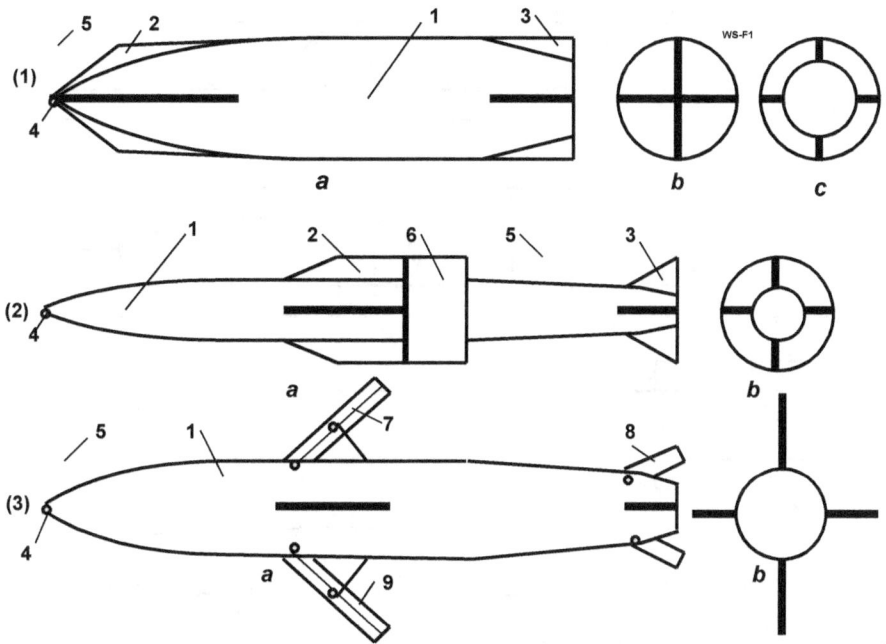

Fig. 2. Types of offered wing projectiles. (1) Full-caliber wing projectile for the rifled and non-rifled guns. (2) Reduced diameter wing projectile having discarding sabot for the rifled and non-rifled guns. (3) Full-caliber projectile with the pull-out mobile variable sweep wings for rifled and non-rifled guns. *a* – side view, *b* – forward view, *c* – back view. Notations: 1 – projectile; 2 – wings; 3 – stabilizer; 4 – optical sensor or navigation system (for example: TV, GPS, laser) for guidance and control (option); 5 – gun barrel; 6 – sabot; 7 – wing; 8 – stabilizer; 9 – flaps (control).

The most guns and rifles have a rifled barrel which rotates the projectile in flight (for stability projectile). That produces the enormous problems for projectile guidance and control. If we do not need a large measure of accuracy (for example the shooting in small village or town) all long distance simplest versions of fig. 2-(1) (without guidance and control) are sufficient. Author made innovations which allow the rotated projectile to create the lift force and have a big range.

If we need in more accuracy, the author offers for rifled gun the projectiles having light free rotated ring which closes the rifled canals and does not allows the powder gas free escape through the rifled cannels (fig.3 – (6)).

In this case the offered innovation allows using the rifled gun as the smoothbore gun. The projectile is thus not forced to rotate.

The other innovation is the special powder cartridge (fig. 3a - (4)). In the conventional patron the gases have a speed limited by the speed of sound: the bullet/projectile cannot reach the speed more 1000 – 2000 m/s in any long barrel. In this proprietary design, the special powder cartridge (4) is placed inside the patron between the bottom of patron and bullet/shell. This cartridge connects to the projectile. In this case the part of powder (into cartridge) will be accelerated together with projectile and pressure (acceleration) will be high pushed by the end of barrel (rocket effect, fig.3b). The projectile speed will be significantly more. In addition the longer the gun barrel, the longer the length of cartridge (fig. 3c) and significantly increases the speed of the projectile. After shooting the cartridge is discarded (fig. 3d).

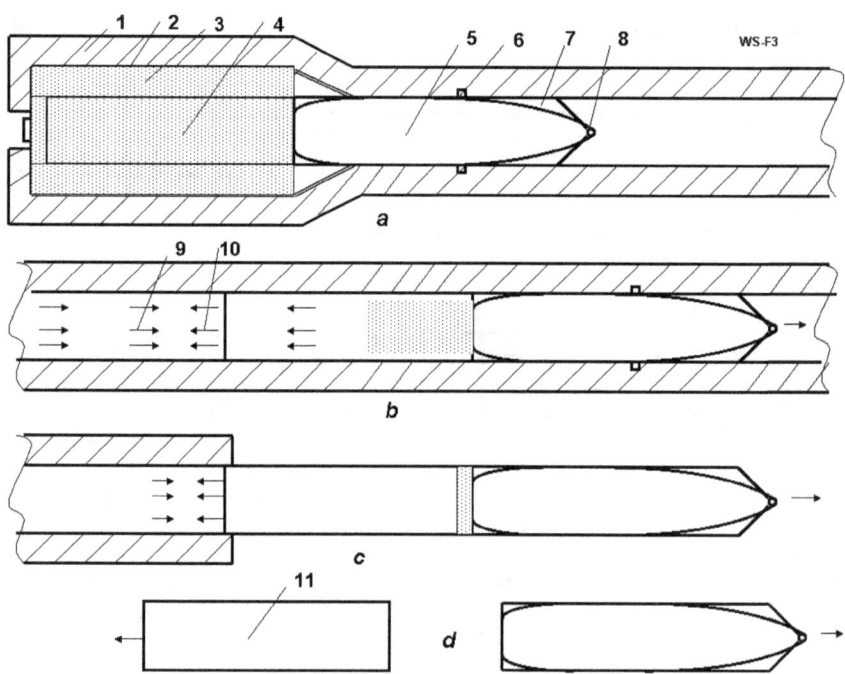

Fig. 3. Wing Projectile with mobile internal cartridge. *a* – initial position, *b* – mobile position inside barrel; *c* – position in the barrel exit, *d* – position out of barrel. Notation: 1 – barrel; 2 – projectile patron; 3 – gun powder out of cartridge; 4 – powder inside of cartridge; 5 – projectile; 6 – light free rotated sealing O-ring (for rifled gun); 7 – wing; 8 – navigation system; 9 – gun gas; 10 – cartridge gas; 11 – cartridge separated from projectile in out of barrel.

The shooting from offered gun is shown in fig. 4. If distance is very long, the projectile launches in top of trajectory the micro-transmitter 9 having small parachute. In moment of explosion the projectile launches the TV transmitter 10 (fig.4), which is transmitting the TV image (11) (result of shooting). List of some *innovations:*

1. Special forms of bullet/shell and wings which increase the range of projectile in 2 – 5 times.
2. The light free rotated sealing O-ring (for rifled gun).
3. Additional special cartridge inside of patron which significantly increases the barrel speed of projectile.
4. Guidance and recognizing of a target image.
5. TV transmitting of shot results.

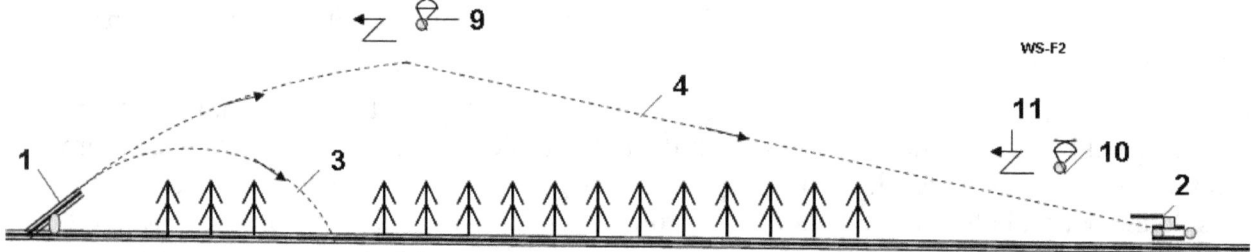

Fig. 4. Shooting at distant targets. Notation: 1 – gun; 2 – long distance target; 3 – conventional short distance trajectory of the non-wing projectile; 4 – long distance trajectory of wing projectile; 9. 10 – TV transmitter; 11 – signal of transmitter (image of result) .

Advantages:

1. Increasing the range in 2 – 5 times.

2. No any changes in guns/rifles. The change is ONLY in projectile (bullet/shell or patron).
3. Using any rifled or non-rifled guns.
4. New type of projectile guidance (recognizing of target).
5. Transmitting of shot result.

3. Theory of flight bullet/shell and a general estimation of range (In metric system)

1. The maximum range, R, of flight bullet/shell is obtained from the kinetic energy of theoretical mechanics for ratio lift/drag $K = $ const. It is equals

$$d\left(\frac{mV^2}{2}\right) = \frac{mg}{K}dR, \quad g = g_0 - \frac{V^2}{R_0}, \quad R = -\frac{KR_0}{2g_0}\ln\frac{g_0 - V_1^2/R_0}{g_0 - V_0^2/R_0}, \quad R \approx \frac{K}{2g}\left(V_1^2 - V_0^2\right), \quad (1)$$

where R is range [m]; $R_0 = 6{,}378 \cdot 10^6$ is the Earth's radius [m]; K is the average aerodynamic efficiency ($K = 6$–18 for subsonic bullet/shell and $K = 2$–5 for supersonic bullet/shell; $g_0 = 9.81$ m/s² is gravity; V_1 is muzzle speed of projectile [m/s]; $V_0 < V_1$ is final (near aim) speed [m/s] ($V_0 = 40$–60 m/s) of projectile; V is variable speed, $V_0 < V < V_1$ [m/s]. For estimation average $V = 0.5(V_1+V_0)$; $mg/K = D$ is air drag [N]; m is bullet/shell mass [kg]. For $V < 2000$ m/s, variable gravity $g \approx g_0$. Last equation in (1) is obtained from the first equation using integration.

The ratio K approximately equals:

$$\text{For } M < 0.9 \quad K \approx 0.5(\pi A/C_{d,0})^{0.5}, \quad \text{where} \quad A = L^2/S,$$
$$\text{For } M > 1{,}5 \quad K \approx 4(1 + 3/M), \quad \text{where} \quad M = V/a, \qquad (2)$$

Here M is Mach number; L is wing span, m; S is wing area, m²; a is sound speed, at $H = 0$, $T = 0°C$ $a = 330$ m/s; for $T = 20°C$ $a = 342$ m/s. For $H > 11$ km $a \approx 295$ m/s; $C_{d,0}$ is the projectile drag coefficient for attack angle $= 0$.

Results of computations for subsonic ($V < 300$ m/s, $M < 0.9$, M is Mach number) and supersonic vehicles are presented in Figs. 5 and 6. The range of a subsonic shell is 30–60 km for $V_1 = 300$ m/s (fig.5); the range of a supersonic shell can reach 400–1000 km for $V_1 = 2000$ m/s (fig.6).

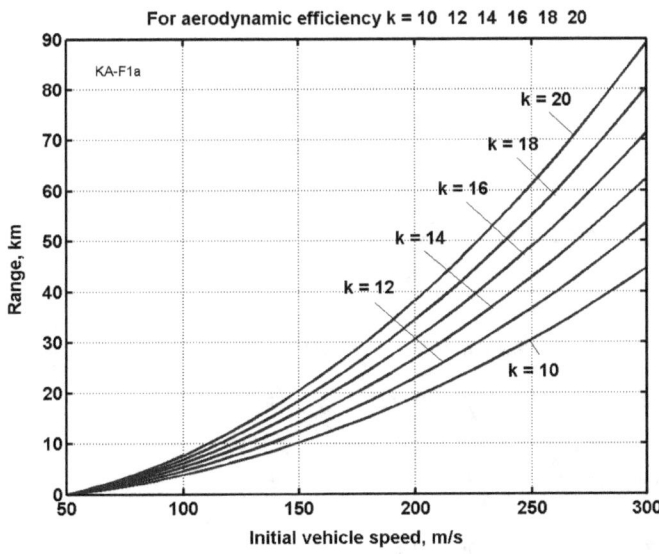

Fig. 5. Range of the subsonic projectile versus initial speed for different aerodynamic efficiency $K = 4 - 16$.

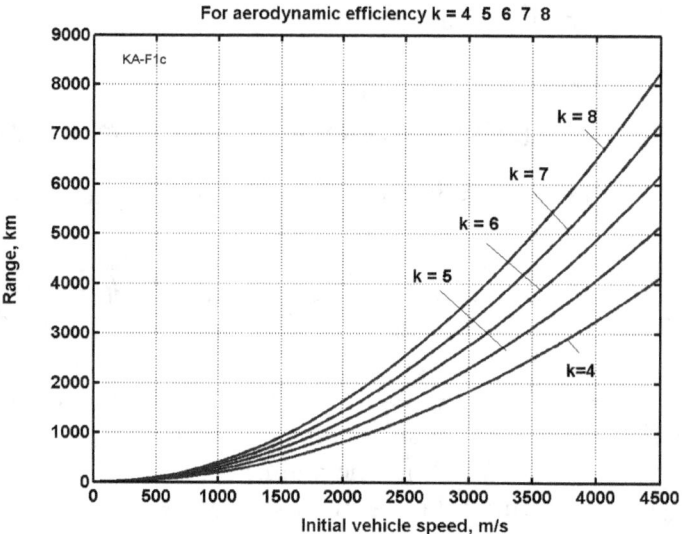

Fig. 6. Range of the supersonic projectile versus initial speed for different aerodynamic efficiency $K = 2\ 3\ 4\ 5$.

2. Average speed and flight time are

$$V_a = \frac{V_1 + V_0}{2}, \quad T = \frac{R}{V_a}. \tag{3}$$

3. Computation of the complex trajectory used the high altitude.

Accuracy equations of ballistic trajectory are:

$$\begin{aligned}
\dot{x} &= V \cos\theta, \\
\dot{H} &= V \sin\theta, \\
\dot{V} &= -\frac{D}{m} - g \sin\theta, \\
\dot{\theta} &= \frac{L}{mV} - \frac{g}{V} \cos\theta.
\end{aligned} \tag{4}$$

For subsonic speed ($M < 0.9$)

$$L = 0.5 C_L \rho V^2 d^2, \quad D = 0.5 C_D \rho V^2 d^2, \tag{5}$$

where

$C_D = C_{d0} + C_{d,w}$, $\quad C_{d,w} = C_{d,w,0} + C_{d,w,i}$, $\quad C_{d,w,0} \approx 0.01 \cdot S / d^2$,

$\lambda = l^2 / S$, $\quad b = 6.25\lambda/(\lambda + 2)$, $\quad \alpha = \theta_o - \theta$, $\quad C_L = b\alpha$, $\quad C_{d,w,i} = C_L / \pi\lambda$.

For supersonic and hypersonic speed ($M > 1.5$)

$$L = 0.5 C_L \rho a V d^2, \quad D = 0.5 C_D \rho a V d^2, \tag{6}$$

$$C_D = C_{d0} + C_{d,w}, \quad C_{d,w} = C_{d,w,0} + C_{d,w,i}, \quad C_{d,w,0} \approx 0.01 \cdot \bar{c}^2,$$
$$\lambda = l^2/S, \quad b = 6.25\lambda/(\lambda+2), \quad \alpha = \theta_o - \theta, \quad C_L = b\alpha, \quad C_{d,w,i} = C_L/\pi\lambda. \tag{6'}$$

where r is range of projectile flight, m; V is projectile speed, m/s; H is projectile altitude, m; θ is trajectory angle, radians; D is projectile drag, N; m is projectile mass, kg; g is gravity at altitude H, m/s^2; L is projectile lift force, N; t is flight time, sec.; C_L is lift force coefficient, for subsonic speed $C_L = 0 - 3.5$, for supersonic speed $C_L \approx 4\alpha$, where α is the wing attack angle, rad; C_D is air drag coefficient. For supersonic wing $C_D \approx \alpha^2$; $a \approx 295$ m/s for $H > 11$ km is sonic speed in atmosphere; S is wing area, m^2; ρ is the air density, for $H = 0$ $\rho_o = 1.225$ kg/m^3. For $H = 0 - 100$ km $\rho \approx \rho_o \exp(-1.4 \cdot 10^{-4})$. We take $C_{d0} = 0.136$ for M < 0.9 and $C_{d0} = 0.473$ for M > 1.2. $\bar{c}$ is relative thickness of wing $\approx 0/05 - 0.1$.

4. Projects

Results of computations are presented below for the different shells and bullets and in Table 1. No optimization of range.

Table 1. Results of computation.

Type of Gun	Type of shell, W- is wing	Caliber mm	Mass of shell. kg	Angle of sight, degree °	Relative wing area	Initial speed, m/s	Final speed, m/s	Range, km	Flight time, sec	Ratio Lift/drag for M<0.9 or M>1.2	Number of Fig.
Rifle	Non W	7.62	0.01	5	0	860	170	2.361	7.84	K=0	Fig.7-1
	Wing	7.62	0.01	5	0.05	860	49	6.852	78	K=4	Fig.7-2
	Wing	6/7.62	0.01	5	0.05	860	62	9.718	88	K=4	Fig.7-3
Rifle	Non W	7.62	0.01	30	0	860	121	4.062	28	K=0	Fig.7-1
	Wing	7.62	0.01	30	0.5	860	33.5	9.890	220	K=4	Fig.7-2
	Wing	6/7.62	0.01	30	0.5	860	42.6	14.10	243	K=4	Fig.7-3
M109	Non W	107	15	30	0.3	494	253	9.432	36.8	K=0	Fig.8-1
	Wing	107	15	30	0.3	494	91	40	354	K=8	Fig.8-2
M107	Non W	155	44	30	0.3	600	314	15.64	48.6	K=0	Fig.9-1
	Wing	155	44	30	0.3	600	108	66	463	K=8	Fig.9-2
Gun	Non W	406	1000	30	0.3	800	445	34.9	71.2	K=0	Fig.10-1
	Wing	406	1000	30	0.3	800	199	140	484	K=8	Fig.10-2
M168	Non W	20	0.102	30	0.3	1050	149	5.896	33.8	K=0	Fig.11-1
	Wing	20	0.102	30	0.3	1050	40	24.9	500	K=8	Fig.11-2
Anti-tank	Non W	84	6.7	8	0.3	290	240	2.088	8	K=0	Fig.12-1
Anti-tank	Wing	84	6.7	8	0.3	290	131	8.636	47.1	K=4	Fig.12-2

Rifle

Mass of shell. kg	Angle of sight, degree °	Relative wing area	Initial speed, m/s
0.01	5	0.05	860

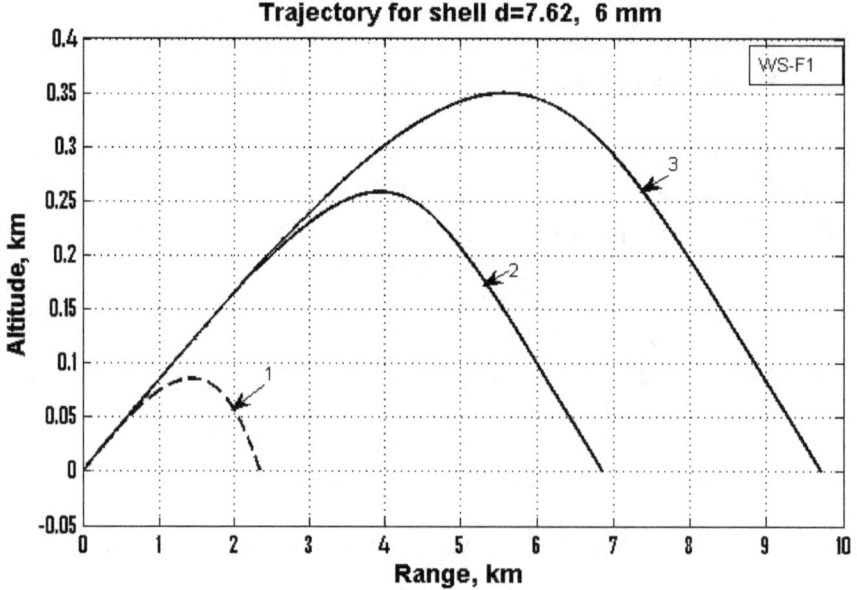

Fig.7a. Rifle $d = 7.62$ mm and 6/7.62. $1 - d = 7.62$ mm, non Wing; $2 - d = 7.62$ mm, Wing; $d = 6/7.62$ mm, Wing, 6 is sub-caliber.

Mass of shell. kg	Angle of sight, degree °	Relative wing area	Initial speed, m/s
0.01	30	0.05	860

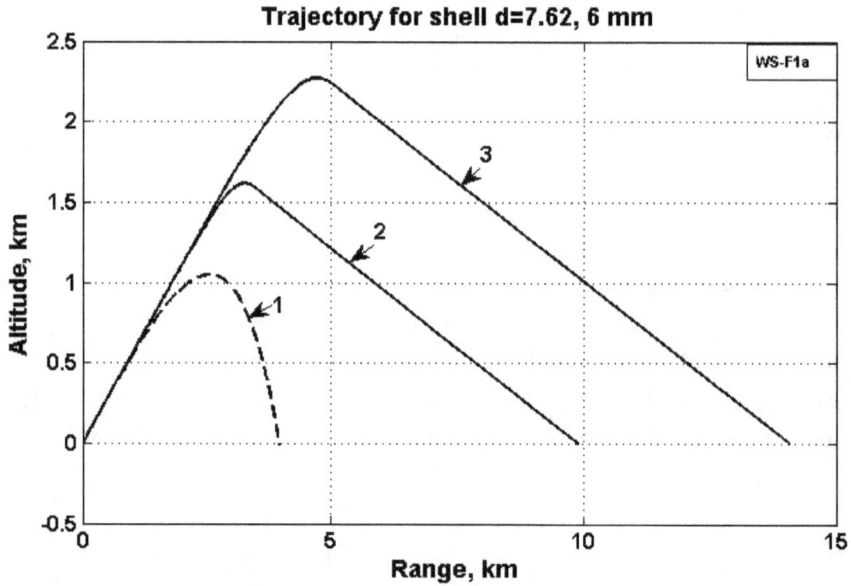

Fig. 7b. Rifle $d = 7.62$ mm and 6/7.62. $1 - d = 7.62$ mm, non Wing; $2 - d = 7.62$ mm, Wing; $d = 6/7.62$ mm, Wing; 6 is sub-caliber.

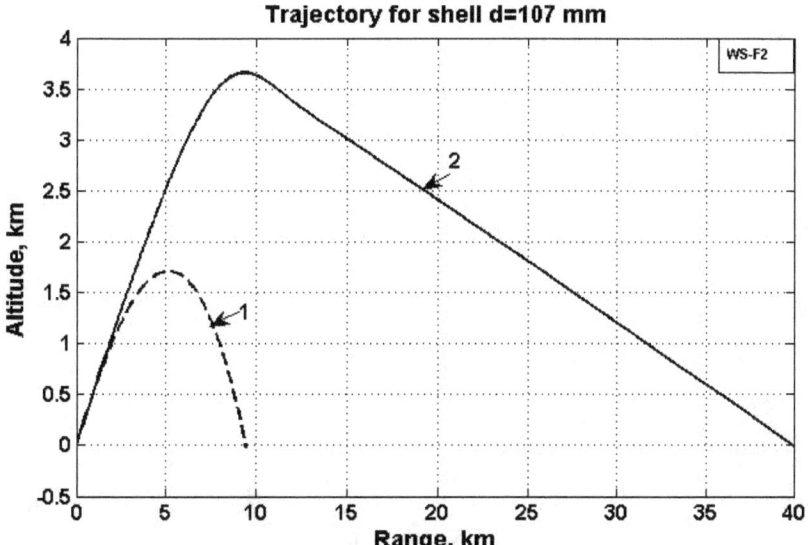

Fig. 8. Howitzer M109, $d = 107$ mm, $M = 15$ kg, $\theta = 30°$, $V_o = 494$ m/s, $S = 0.3$, $K = 8$.
1 – Conventional shell; 2 – Shell has wing.

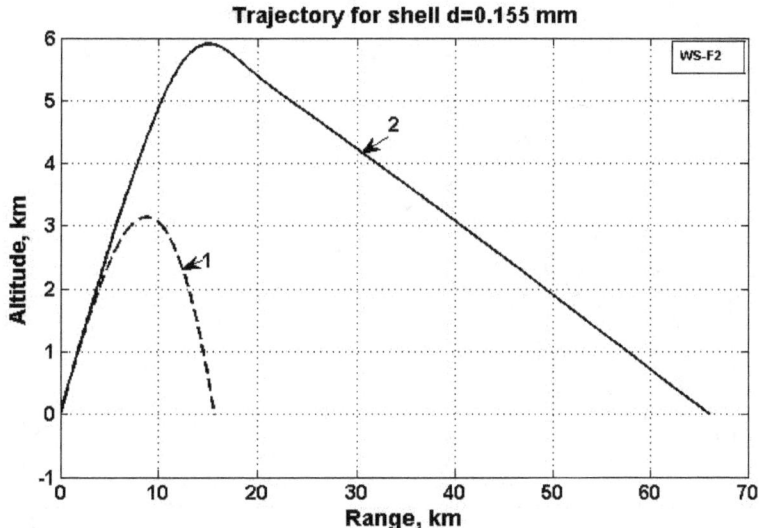

Fig. 9. Howitzer M107 $d = 155$ mm, $M = 44$ kg, $\theta = 30°$, $V_o = 600$ m/s, $S = 0.3$, $K = 8$.
1 – Conventional shell; 2 – Shell has wing

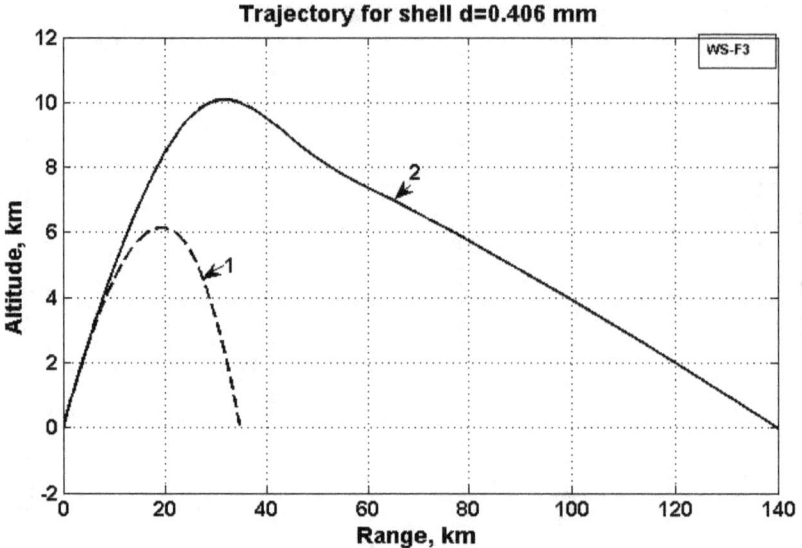

Fig. 10. Big warship gun, $d = 406$ mm, $M = 1000$ kg, $\theta = 30°$,
$V_o = 800$ m/s, $S = 0.3$, $K = 8$
1 – Conventional shell; 2 – Shell has wing

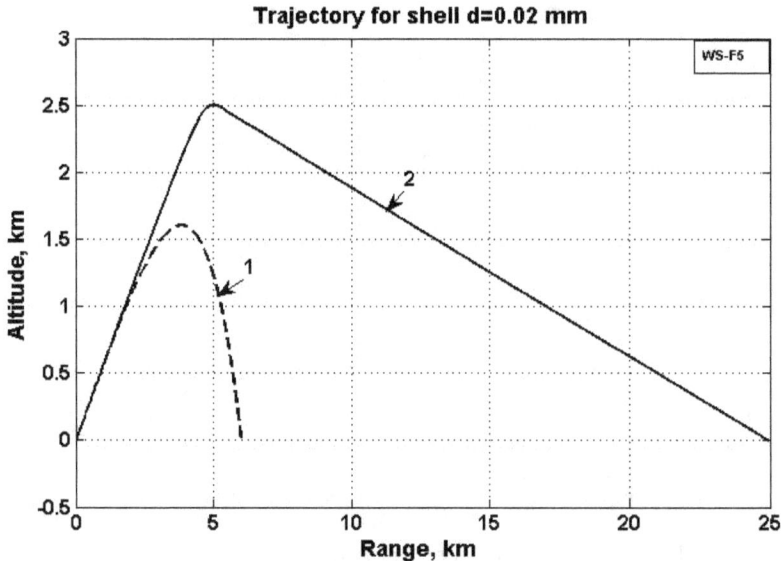

Fig. 11. Anti-aircraft gun M168, $d = 20$ mm,
$M = 0.1$ kg, $\theta = 30°$, $V_o = 1050$ m/s, $S = 0.3$, $K = 8$.
1 – Conventional shell; 2 – Shell has wing

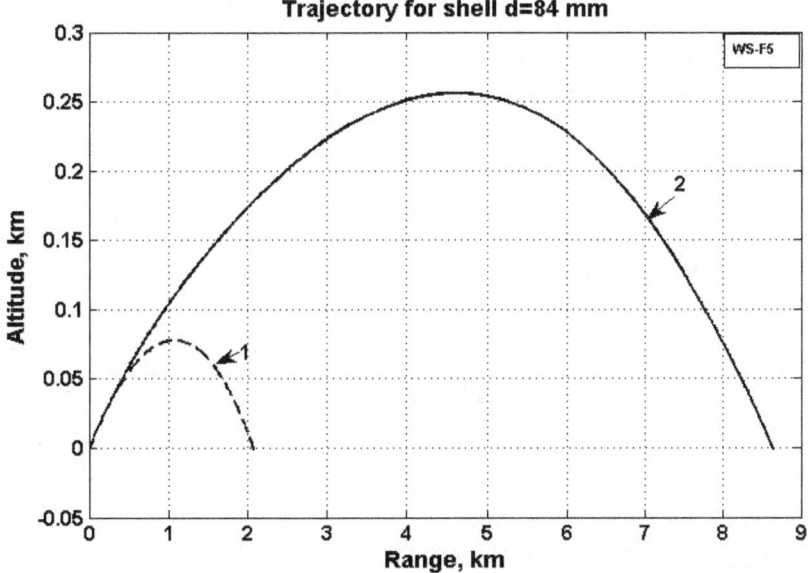

Fig.12. Anti-tank, $d = 84$ mm, $M = 6{,}7$ kg, $\theta = 8°$, $V_0 = 290$ m/s, $S = 0.3$, $K = 4$. 1 – Conventional shell; 2 – Shell has wing.

5. Conclusion

New forms of shells/bullets which increases range of the projectiles 2 – 5 times is described. These forms contain in its design small special wings and stabilizers. The shell/bullet special form wings support the projectile in air, so that unlike conventional bullets or shells at that distance, these do fall to earth's surface and the projectile maintains significant kinetic energy. The important innovation is its compatibility with the conventional rifles and gun with rifled barrel.

The second idea is radical change of trajectory. The projectile reaches a high altitude and glides from height using wings with subsonic speed and a good ratio lift/drag.

Author developed the theory of these projectiles and computed some projects which show high efficiency of this innovations. These bullets and shells can be quickly integrated into the arms industry and army because it does not require manufacture of new weapons (rifles, guns), but only change the bullets and shells.

Author also suggests for this systems: the light free rotated sealing O-ring (for rifled gun); additional special cartridge inside of patron which significantly increases the barrel speed of projectile; guidance by recognizing of a target image; TV transmitting of shot results.
The additional information about this topic are [8]-[14].

Acknowledgement

The author wishes to acknowledge Shmuel Neumann (Israel) for correcting the English and offering useful advices and suggestions.

References:

1. Bolonkin A.A., (1972). *New Methods of Optimization and Their Applications*, Moscow, MHTU, 1972, 220 ps. (in Russian).

2. Bolonkin A.A., (1994). Methods of Optimization, Work presented to *AIAA/NASA/USAF/SSMO Symposium on Multidisciplinary Analysis and Optimization*, 7 – 9 September 1994, Panama City, FL, USA.
3. Bolonkin A.A., (2004). Optimal Trajectories of Air and Space Vehicles. Journal *"Aircraft Engineering and Space Technology"*. Vo.76, No.2, 2004, pp.193-214.
4. Bolonkin A.A., (2006). *Non-Rocket Space Launch and Flight*. Elsevier, 2006, 468 ps. Attn. 4, pp.368-423. http://www.archive.org/details/Non-rocketSpaceLaunchAndFlight , http://www.scribd.com/doc/24056182 .
5. Bolonkin A.A.,(2006). *"New Concepts, Ideas, Innovations in Aerospace, Technology and the Human Sciences"*, NOVA, 2006, 510 pgs. http://www.scribd.com/doc/24057071, http://www.archive.org/details/NewConceptsIfeasAndInnovationsInAerospaceTechnologyAndHumanSciences.
6. Bolonkin A.A., Cathcart R.B.,(2007). *"Macro-Projects: Environments and Technologies"*, NOVA, 2007, 536 pgs. http://www.scribd.com/doc/24057930 . http://www.archive.org/details/Macro-projectsEnvironmentsAndTechnologies
7. Bolonkin A.A., (2011). *Femtotechnologies and Revolutionary Projects*. Scribd, USA, 2011. 538 p. 16 Mb. http://www.scribd.com/doc/75519828/ http://www.archive.org/details/FemtotechnologiesAndRevolutionaryProjects
8. Evans, Nigel F. (2007) "British Artillery in World War 2
9. Excalibur XM982 - Defense Update, (2012) http://www.defense-update.com/products/e/excalibur.htm .
10. 2012 RAND report (2012), http://www.rand.org/pubs/monographs/MG1171z2.html
11. F.D. Witherspoon, (2009), "MiniRailgun accelerator for plasma linear driven HEDP and magneto-inertial fusion experiments," IEEE Int. Conf. Plasma Sci., San Diego, CA, 2009.
12. McNab, Chris; Hunter Keeter (2008). *Tools of Violence: Guns, Tanks and Dirty Bombs*. Osprey Publishing. ISBN 1-84603-225-3
13. John Pike (2012). "XM982 Excalibur 155mm Precision Guided Extended Range Artillery Projectile". Globalsecurity.org. http://www.globalsecurity.org/military/systems/munitions/m982-155-program.htm. Retrieved 2012-07-06.
14. Clark, Colin (2010-10-27). "Excalibur Use Rises In Afghanistan". DoD Buzz. http://www.dodbuzz.com/2010/10/27/excalibur-use-rises-in-afghanistan/. Retrieved 2012-07-06.

October 2012.

Chapter 15

Deep Penemration Bomb

Abstract

Authors offer the new anti-bunker bombs which reach 80-150 m and more of the Earth depth. They can destroy armor protected underground bunkers. This bomb is named as "Self-propelled Bomb" because after conventional kinetic penetration, multiple cumulative charges creates a narrow canal, then injects into this canal explosives which upon detonation pushes the bomb deeper into the Earth by special rocket explosions and reaches a deep location. The other feature of Burn Bomb is the use of liquid explosive which makes it more comfortable, easy for design, safety and operates more effective than current bunker buster bomb. The same method may be used for super-fast very deep oil/gas drilling because the liquid explosive may be delivered to same apparatus by a long tube line.

Introduction

Inadequacy to Terminate Iran Nuclear Weapons Program

Despite the intolerable threat of a nuclear Iran, the United States appears to lack the technology to inflict severe damage on Iran's atomic weapons program. According to the Wall Street Journal and other reliable sources, the United States simply does not have weapons to destroy enough of the right targets that, if damaged or destroyed would significantly slow or stop Iran's weapons program. It is more than likely that the west has imperfect knowledge about Iranian atomic facilities, especially those with a weapons nexus. However, the following are the known sites which would have to be targeted.

Esfahan is an above ground uranium conversion facility that converts raw material into uranium gas which is then shipped to the Natanz facility for enrichment. The complex includes an extensive tunnel complex which could house more sensitive uranium activities.

Natanz is an underground enrichment facility buried under 25 meters of earth with a 2.5-meter thick concrete ceiling and houses at least 8,000 centrifuges which have turned out enough material for several nuclear warheads. The complex includes three large underground buildings, two of which are designed to be cascade halls to hold 50,000 centrifuges.

Fordow is an underground enrichment facility buried 80 meters inside a mountain and protected by anti-aircraft weapons. Recently uranium fuel arrived for further enrichment. The facility is large and safe enough from attack to provide for quick weapons grade enrichment.

Arak is a heavy water production plant. The above ground plant once operational could produce about 9 kilograms of plutonium annually or enough for about two nuclear weapons.

Bushehr is an above ground 1,000-megawatt reactor. The fuel from this facility is sufficient to produce 50 to 75 bombs.

Parchin is a high explosives testing site which houses a containment vessel used to conduct tests of the high explosives used in triggering a fissile reaction.

Mojdeh is the center for weapons development located on the Ministry of Defense's Malek-Ashtar University of Technology in Esfahan. It works on the trigger for an atomic bomb, casting and machining of uranium metals, research on fissile material needed for a bomb, high explosives and radiation detection.

Abyek, a formerly top secret nuclear site is inside a mountain and has three large halls, 20 by 200 meters, and is **100 meters** below the mountain surface. It is one of the newest command centers under the direction of Mojdeh. It is noteworthy that in 2010 Tehran announced plans to build 10 additional enrichment sites inside mountains beginning in March 2011. It appears Abyek is the first of those sites.

State of the Art of Bunker Busters

These targets vary in vulnerability. The above ground unfortified facilities are easy targets for standoff cruise missiles but the hard and deeply buried targets (HDBT) are especially challenging. U.S.-made bunker-buster bombs for HDBT might breech the cavity containing some of Iran's buried facilities. The GBU-27 can penetrate 2.4 meters of concrete and the GBU-28 can penetrate 6 meters of concrete and another layer of earth 30 meters deep. Last week, the Washington, DC-based Bipartisan Policy Center's National Security Project called for providing Israel 200 GBU-31 bombs, which include the Boeing Co. GPS tail-kit, to increase the credibility of a strike. An article in Israel's Tablet magazine suggested Israel might attack HDBT sites like Fordow with a series of bunker busters, dropped at the same point to burrow through the granite. Successfully striking an HDBT depends on accuracy of fuse settings which depends on knowing with great accuracy the types of cover, such as the PSI of the concrete, types of layering, and depth. The most accurate fuses rely on delays, and the delay settings are determined by the time it takes for the weapon to travel from impact to the area of detonation.

The greatest limitation of the enormous penetration bomb GBU-57A/B is that this bomb is very heavy (14 tons) and as such, must be delivered only by large bombers. Worse, this bunker buster bomb claims to be effective in destroying a bunker located underground 60 m. This exaggerated claim is probably part of a necessary disinformation campaign, and in reality this bomb's effective depth is more like 30 meters. Even worse, it is very likely that the underground nuclear facilities are armored, not just by the commonly known thick layer of steel reinforced concrete.

An example of a Russian bunker buster is the KAB-1500L-Pr. It is delivered with the Su-24M and the Su-27IB aircraft. It is claimed to be able to penetrate 10-20 m of earth or 2 m of reinforced concrete. The bomb weighs 1,500 kg (3,300 lb.), with 1,100 kg (2,400 lb.) being the high explosive penetrating warhead. It is laser guided and has a reported strike accuracy of 7 m (23 ft.) CEP.

The US has a series of custom made bombs to penetrate hardened or deeply buried structures:

Depth of Penetration	Weapon Systems	
Penetration of reinforced concrete: 1.8 m (6 ft.)	BLU-109 Penetrator	GBU-10, GBU-15, GBU-24, GBU-27, AGM-130
Penetration of reinforced concrete: 3.4 m (11 ft.)	BLU-116 Advanced Unitary Penetrator (AUP)	GBU-15, GBU-24, GBU-27, AGM-130
	BLU-118/B Thermobaric Warhead	GBU-15, GBU-24, AGM-130

Penetration of reinforced concrete: more than 6 m (20 ft.) BLU-113 Super Penetrator   GBU-28, GBU-37

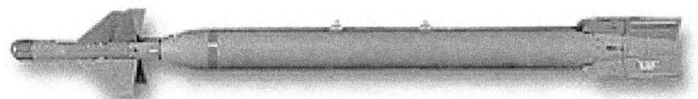

Bomb GBU-28

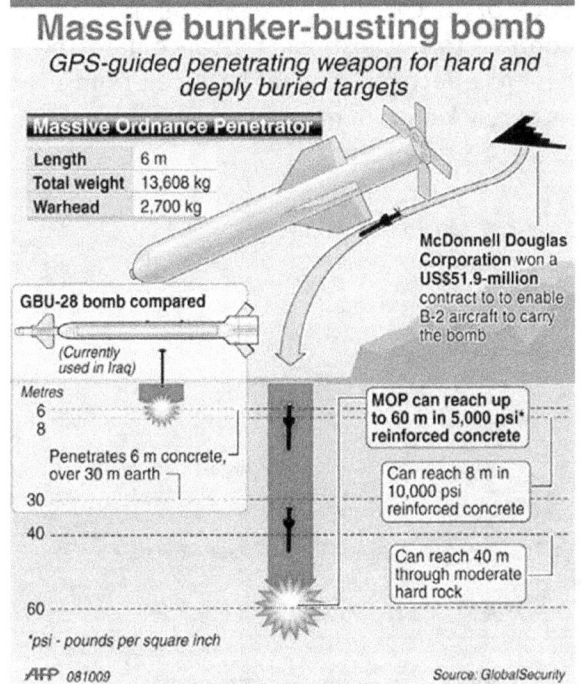

Fig.1. Penetration bomb GBU-28 (left) and GBU-57 (right). For 5000 psi the penetration of GBU-28 and CBU-57 is same.

More recently, the US has developed the 30,000-pound GBU-57. The Massive Ordnance Penetrator (MOP) GBU-57A/B is a U.S. Air Force massive, precision-guided, 30,000-pound (13,608 kg) "bunker buster" bomb. This is substantially larger than the deepest penetrating bunker busters previously available, the 5,000-pound (2,268 kg) GBU-28 and GBU-37.

The need for greater penetration bombs became salient following the 2003 invasion of Iraq, in which analysis of sites that had been targeted with bunker-buster bombs revealed poor penetration and inadequate levels of destruction. This renewed interest in the development of a super-large bunker-buster, and the MOP project was initiated by the Defense Threat Reduction Agency to fulfill a long-standing Air Force requirement. The U.S. Air Force has a call for a collection of massively sized penetrator and blast weapons, the so-called "Big BLU" collection, which includes the MOAB (Massive Ordnance Air Burst) bomb. Development of the MOP is now underway at the Air Force Research Laboratory, Munitions Directorate, Eglin Air Force Base, Florida. Design and testing work is also being performed by Boeing. The initial explosive test of MOP took place on March 14, 2007 in a tunnel belonging to the Defense Threat Reduction Agency (DTRA) at the White Sands Missile Range, New Mexico. The project has had at least one successful Flight Test MOP launch. The final testing will be completed in 2012. The Air Force took delivery of 20 bombs, designed to be delivered by the B-2 bomber, in September 2011. In February 2012, Congress approved $81.6 million to further develop and improve the weapon.

Mechanics of Penetration Bombs

Penetration bombs use kinetic energy and sometimes a shaped charge, an explosive charge shaped to focus the effect of the explosive's energy. Various types are used to cut and form metal, to initiate nuclear weapons, to penetrate armor, and to "complete" wells in the oil and gas industry. A typical modern lined shaped charge can penetrate armor steel to a depth of 7 or more times the diameter of the charge (charge diameters, CD), though greater depths of 10 CD and above have been achieved. Contrary to a widespread misconception, the shaped charge does not depend in any way on heating or melting for its effectiveness, that is, the jet from a shaped charge does not melt its way through armor, as its effect is purely kinetic in nature.

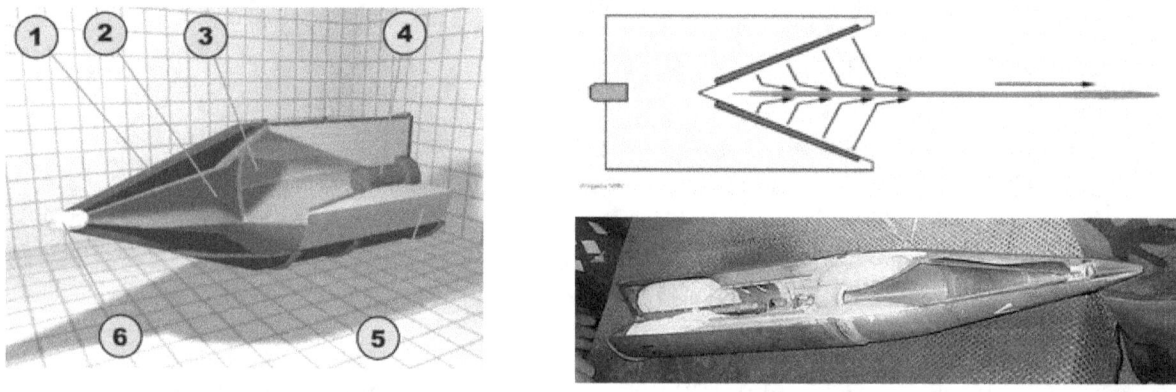

Figure 2 (left): Sections 1: Aerodynamic cover; 2: Empty cavity; 3: Conical liner; 4: Detonator; 5: Explosive; 6: Piezo-electric trigger
Figure 3 (Top-right). Work of shaped charge.
Figure 4 (bottom-right): Sectioned high explosive anti-tank round with the inner shaped charge visible

A typical device consists of a solid cylinder of explosive with a metal-lined conical hollow in one end and a central detonator, array of detonators, or detonation wave guide at the other end. Explosive energy is released directly away from (normal to) the surface of an explosive, so shaping the explosive will concentrate the explosive energy in the void. If the hollow is properly shaped (usually conically), the enormous pressure generated by the detonation of the explosive drives the liner in the hollow cavity inward to collapse upon its central axis. The resulting collision forms and projects a high-velocity jet of metal forward along the axis. Most of the jet material originates from the innermost part of the liner, a layer of about 10% to 20% of the thickness. The rest of the liner forms a slower-moving slug of material, which, because of its appearance, is sometimes called a "carrot".

Because of the variation along the liner in its collapse velocity, the jet's velocity also varies along its length, decreasing from the front. This variation in jet velocity stretches it and eventually leads to its break-up into particles. Over time, the particles tend to fall out of alignment, which reduces the depth of penetration at long standoffs.

Also, at the apex of the cone, which forms the very front of the jet, the liner does not have time to be fully accelerated before it forms its part of the jet. This results in its small part of jet being projected at a lower velocity than the jet formed later behind it. As a result, the initial parts of the jet coalesce to form a pronounced wider tip portion.

Most of the jet travels at hypersonic speed. The tip moves at 7 to 14 km/s, the jet tail at a lower velocity (1 to 3 km/s), and the slug at a still lower velocity (less than 1 km/s). The exact velocities depend on the charge's configuration and confinement, explosive type, materials used, and the explosive-initiation mode. At typical velocities, the penetration process generates such enormous pressures that it may be considered hydrodynamic; to a good approximation, the jet and armor may be treated as inviscid, incompressible fluid, with their material strengths ignored.

The location of the charge relative to its target is critical for optimum penetration, for two reasons. If the charge is detonated too close there is not enough time for the jet to fully develop. But the jet disintegrates and disperses after a relatively short distance, usually well under 2 meters. At such standoffs, it breaks into particles which tend to tumble and drift off the axis of penetration, so that the successive particles tend to widen rather than deepen the hole. At very long standoffs, velocity is lost to air drag, further degrading penetration.

The key to the effectiveness of the hollow charge is its diameter. As the penetration continues through the target, the width of the hole decreases leading to a characteristic "fist to finger" action, where the size of the eventual "finger" is based on the size of the original "fist". In general, shaped charges can penetrate a steel plate as thick as 150% to 700% of their diameter, depending on the charge quality. The figure is for basic steel plate, not for the composite armor, reactive armor, or other types of modern armor.

The Explosive

For optimal penetration, a high explosive having a high detonation velocity and pressure is normally chosen. The most common explosive used in high performance anti-armor warheads is HMX (octogen), though it is never used in pure form, as it would be too sensitive. It is normally compounded with a few percent of some type of plastic binder, such as in the polymer-bonded explosive (PBX) LX-14, or with another less-sensitive explosive, such as TNT, with which it forms Octol. Other common high-performance explosives are RDX-based compositions, again either as PBXs or mixtures with TNT (to form Composition B and the Cyclotols) or wax (Cyclonites). Some explosives incorporate powdered aluminum to increase their blast and detonation temperature, but this addition generally results in decreased performance of the shaped charge. There has been research into using the very high-performance but sensitive explosive CL-20 in shaped-charge warheads, but, at present, due to its sensitivity, this has been in the form of the PBX composite LX-19 (CL-20 and Estane binder).

Other Features

A _waveshaper_ is a body (typically a disc or cylindrical block) of an inert material (typically solid or foamed plastic, but sometimes metal, perhaps hollow) inserted within the explosive for the purpose of changing the path of the detonation wave. The effect is to modify the collapse of the cone and resulting jet formation, with the intent of increasing penetration performance. Waveshapers are often used to save space; a shorter charge can achieve the same performance as a longer one without a waveshaper.

Another useful design feature is *sub-calibration*, the use of a liner having a smaller diameter (caliber) than the explosive charge. In an ordinary charge, the explosive near the base of the cone is so thin that it is unable to accelerate the adjacent liner to sufficient velocity to form an effective jet. In a sub-calibrated charge, this part of the device is effectively cut off, resulting in a shorter charge with the same performance.

Shaped Charge Variants

Explosively Formed Penetrator

A conventional shaped charge generally has a conical metal liner that projects a hypervelocity jet of metal able to penetrate to great depths into steel armor; in travel over some distance the jet breaks up along its length into particles that drift out of alignment, greatly diminishing its effectiveness at a distance.

An Explosively Formed Penetrator or EFP, on the other hand, has a liner face in the shape of a shallow dish. The force of the blast molds the liner into any of a number of shapes, depending on the shape of the plate and how the explosive is detonated. Some sophisticated EFP warheads have multiple detonators that can be fired in different arrangements causing different types of waveform in the explosive, resulting in either a long-rod penetrator, an aerodynamic slug projectile, or multiple high-velocity fragments. A less sophisticated approach for changing the formation of an EFP is the use of wire-mesh in front of the liner: with the mesh in place the liner fragments into multiple penetrators.

In addition to single-penetrator EFPs (also called single EFPs or SEFPs), there are EFP warheads whose liners are designed to produce more than one penetrator; these are known as multiple EFPs, or MEFPs. The liner of an MEFP generally comprises a number of dimples that intersect each other at sharp angles. Upon detonation the liner fragments along these intersections to form up to dozens of small, generally spheroidal projectiles, producing an effect similar to that of a shotgun. The pattern of impacts on target can be finely controlled based on the design of the liner and the manner in which the explosive charge is detonated. A nuclear-driven MEFP was apparently proposed by a member of the JASON group in 1966 for terminal ballistic missile defense. A related device was the proposed nuclear pulse propulsion unit for Project Orion.

The (single) EFP generally remains intact and is therefore able to penetrate armor at a long range, delivering a wide spray of fragments of liner material and vehicle armor into the vehicle's interior, injuring its crew and damaging other systems.

As a rule of thumb, an EFP can perforate a thickness of armor steel equal to half the diameter of its charge for a copper or iron liner, and armor steel equal to the diameter of its charge for a tantalum liner, whereas a typical shaped charge will go through six or more diameters.

The penetration is proportional to the density of the liner metal; tantalum 16,654 g/cm3, copper 8,960 g/cm3, iron 7,874 g/cm3. Tantalum is preferable in delivery systems that have limitations in size, like the SADARM, which is delivered by a howitzer. For other weapon systems where the size does not matter, a copper liner of double the caliber is used.

Extensive research is going on in the zone between jetting charges and EFPs, which combines the advantages of both types, resulting in very long stretched-rod EFPs for short-to-medium distances (because of the lack of aero stability) with improved penetration capability.

EFPs have been adopted as warheads in a number of weapon systems, including the CBU-97 and BLU-108 air bombs (with the Skeet submunition), the M303 Special Operations Forces demolition kit, the M2/M4 Selectable Lightweight Attack Munition (SLAM), the SADARM submunition, the Low Cost Autonomous Attack System, and the TOW-2B anti-tank missile.

An EFP eight inches in diameter threw a seven-pound copper slug at Mach 6, or 2,000 meters per second. (A .50-caliber bullet, among the most devastating projectiles on the battlefield, weighs less than two ounces and has a muzzle velocity of 900 meters per second.).— Rick Atkinson, *The Washington Post.*

Self-Forging Projectile

The Explosively Formed Penetrator (EFP) is also known as the Self-Forging Fragment (SFF), Explosively Formed Projectile (EFP), SElf-FOrging Projectile (SEFOP), Plate Charge, and Misznay-Schardin (MS) Charge. An EFP uses the action of the explosive's detonation wave (and to a lesser extent the propulsive effect of its detonation products) to project and deform a plate or dish of ductile metal (such as copper, iron, or tantalum) into a compact high-velocity projectile, commonly called the slug. This slug is projected toward the target at about two kilometers per second. The chief advantage of the EFP over a conventional (e.g., conical) shaped charge is its effectiveness at very great standoffs, equal to hundreds of times the charge's diameter (perhaps a hundred meters for a practical device).

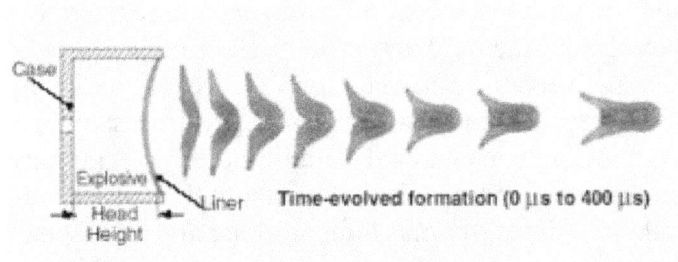

Fig.5. Self-Fording Projectile and formation of slug.

The EFP is relatively unaffected by first-generation reactive armor and can travel up to perhaps 1000 charge diameters (CDs) before its velocity becomes ineffective at penetrating armor due to aerodynamic drag, or successfully hitting the target becomes a problem. The impact of a ball or slug EFP normally causes a large-diameter but relatively shallow hole, of, at most, a couple of CDs. If the EFP perforates the armor, spalling and extensive behind armor effects (BAE, also called behind armor damage, BAD) will occur. The BAE is mainly caused by the high-temperature and high-velocity armor and slug fragments being injected into the interior space and the blast overpressure caused by this debris. More modern EFP warhead versions, through the use of advanced initiation modes, can also produce long-rods (stretched slugs), multi-slugs and finned rod/slug projectiles. The long-rods are able to penetrate a much greater depth of armor, at some loss to BAE, multi-slugs are better at defeating light or area targets and the finned projectiles are much more accurate. The use of this warhead type is mainly restricted to lightly armored areas of main battle tanks (MBT) such as the top, belly and rear armored areas. It is well suited for the attack of other less heavily protected armored fighting vehicles (AFV) and in the breaching of material targets (buildings, bunkers, bridge supports, etc.). The newer rod projectiles may be effective against the more heavily armored areas of MBTs. Weapons using the EFP principle have already been used in combat; the "smart" submunitions in the CBU-97 cluster bomb used by the US Air Force and Navy in the 2003 Iraq war employed this principle, and the US Army is reportedly experimenting with precision-guided artillery shells under Project SADARM (Seek And Destroy ARMor). There are also various other projectile (BONUS, DM 642) and rocket submunitions (Motiv-3M, DM 642) and mines (MIFF, TMRP-6) that use EFP principle. Examples of EFP warheads are US patents 5038683 and US6606951.

Liquid Explosives

One of the innovations of the proposed New Generation Penetration Bomb is the use of liquid rather than solid explosives. Certainly not all liquid explosives are common domain knowledge but some candidates follow.

Oxyliquit

An oxyliquit, also called liquid air explosive or liquid oxygen explosive, is an explosive material made of a mixture of liquid air or liquid oxygen (LOX) with a suitable fuel, usually carbon (as lampblack) or some organic chemical (e.g. a mixture of soot and naphthalene), wood meal, or aluminum powder or sponge; the material is capable of absorbing several times its weight of LOX. It is a class of Sprengel explosives which is a generic class of materials invented by Hermann Sprengel in the 1870s. They consist of stoichiometric mixtures of strong oxidizers and reactive fuels, mixed just prior to use in order to enhance safety. Either the oxidizer or the fuel, or both, should be a liquid to facilitate mixing, and intimate contact between the materials for a fast reaction rate. Sprengel suggested nitric acid, nitrates and chlorates as oxidizers, and nitroaromatics (e.g. nitrobenzene) as fuels. Other Sprengel explosives used at various times include charcoal with liquid oxygen (an oxyliquit), "Rackarock", and ANFO ammonium nitrate (oxidizer) mixed with a fuel oil (fuel), normally diesel kerosene or nitromethane "Rackarock" consisted of potassium chlorate and nitrobenzene. It was provided in the form of permeable cartridges of the chlorate, which were placed in wire baskets and dipped in the nitrobenzene for a few seconds before use. It was famously used in the massive submarine demolition of a navigational hazard in Long Island Sound in 1885. The charge of over a hundred tonnes of explosive (laid in tunnels 20 meters below sea level) destroyed approximately 600,000 tonnes of rock, and created a wave 30 m high.

A mixture of lampblack and liquid oxygen was measured to have detonation velocity of 3,000 m/s, and 4 to 12% more strength than dynamite. However, the flame it produces has too long duration to be safe in possible presence of explosive gases, so oxyliquits found their use mostly in open quarries and strip mining. However, this is a candidate for liquid explosives for the New Generation Penetration

Bomb that may also be used as rocket fuel to propel the bomb to great velocities before impact. As a disadvantage, oxyliquits, once mixed, are sensitive to sparks, shock and friction, and there were reported cases of spontaneous ignition. The power relative to weight is high, but the density is low, so the brisance is low as well. Ignition by a fuse alone is sometimes unreliable. The charge should be detonated within 5 minutes of soaking, but even after 15 minutes it may be capable of exploding, even though weaker and with production of carbon monoxide.

Nitroglycerin and Pentaerythritol tetranitrate (PETN).

The best known liquid explosive, Nitroglycerin is a high explosive which is so unstable that the slightest jolt, friction, or impact can cause it to detonate. The molecule contains oxygen, nitrogen, and carbon with weak chemical bonds. Hence when it explodes, great energy is released as the atoms rearrange to form new molecules with strong, stable bonds such as N2, H2O, and CO2. It is the speed of the decomposition reaction which makes it such a violent explosive. A supersonic wave passing through the material causes it to decompose almost instantly. This is an unlikely candidate because of its instability.

Structurally, PETN (Chemical Abstract Services Registry Number 78-11-5) structurally resembles nitroglycerin, and is also known as PENT, PENTA, TEN, corpent, penthrite, is the nitrate ester of pentaerythritol. PETN is one of the most powerful high explosives known, with a relative effectiveness factor of 1.66. PETN is practically insoluble in water (0.01 g/100 ml at 50 °C), weakly soluble in common nonpolar solvents such as aliphatic hydrocarbons (like gasoline) or tetrachloromethane, but soluble in some other organic solvents, particularly in acetone (about 15 g/100 g of the solution at 20 °C, 55 g/100 g at 60 °C) and dimethylformamide (40 g/100 g of the solution at 40 °C, 70 g/100 g at 70 °C). PETN forms eutectic mixtures with some liquid or molten aromatic nitro compounds, e.g. trinitrotoluene (TNT) or tetryl. Due to its highly symmetrical structure, PETN is resistant to attack by many chemical reagents; it does not hydrolyze in water at room temperature or in weaker alkaline aqueous solutions. Water at 100° or above causes hydrolysis to dinitrate; presence of 0.1% nitric acid accelerates the reaction. Addition of certain aromatic nitro derivatives lowers thermal stability of PETN.

PETN is as an explosive with high brisance and its basic explosion characteristics are:

- Explosion energy: 5810 kJ/kg (1390 kcal/kg), so 1 kg of PETN has the energy of 1.24 kg TNT.
- Detonation velocity: 8350 m/s (1.73 g/cm^3), 7910 m/s (1.62 g/cm^3), 7420 m/s (1.5 g/cm^3), 8500 m/s (pressed in a steel tube)
- Volume of gases produced: 790 dm^3/kg (other value: 768 dm^3/kg)
- Explosion temperature: 4230 °C
- Oxygen balance: -6.31 atom -g/kg
- Melting point: 141.3 °C (pure), 140–141 °C (technical)
- Trauzl lead block test: 523 cm^3 (other values: 500 cm^3 when sealed with sand, or 560 cm^3 when sealed with water)
- Critical diameter (minimal diameter of a rod that can sustain detonation propagation): 0.9 mm for PETN at 1 g/cm^3, smaller for higher densities (other value: 1.5 mm)

Nonphlegmatized PETN is stored and handled with approximately 10% water content. PETN has been replaced in many applications by RDX, which is thermally more stable and has longer shelf life. Replacement of the central carbon atom with silicon produces Si-PETN, which is extremely sensitive. PETN can be initiated by a laser. A pulse with duration of 25 nanoseconds and 0.5–4.2 joules of energy from a Q-switched ruby laser can initiate detonation of a PETN surface coated with a 100 nm thick aluminum layer in less than half microsecond.

Description, Workings and Innovations of new Bomb

Description. The offered penetration bunker bomb (Self-propelled bomb) is shown in Fig.6. One contains: the body 1; forward part 2 (initial implantation); flight accelerator 4; explosion chamber (underground engine) 5; folding hooks 6; main shaped (cumulative) chamber 7; channels for exhaust gas 8; tank for a liquid explosives 9 having one or two component; injectors for liquid explosives 10 – 12.

The forward part 2 contains the initial shaped (cumulative) chamber 3.

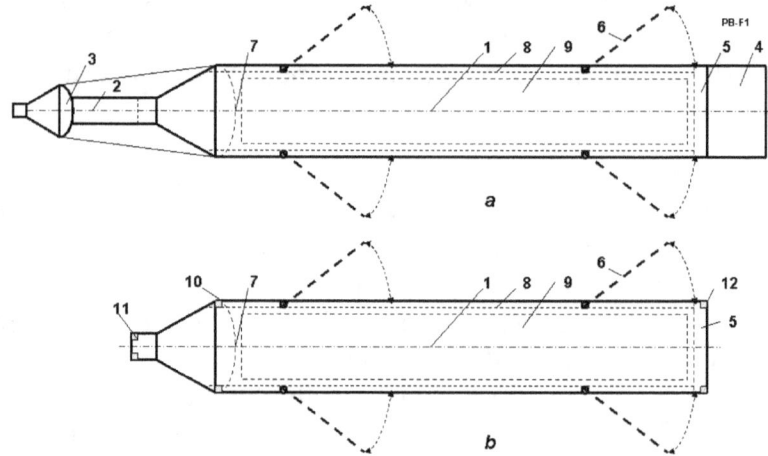

Fig. 6, Suggested penetration bunker bomb (Self-propelled bomb). (a) Flight bomb; (b) Underground part of Self Propelled bomb. Notations: 1 – body of bomb; 2 – Initial implantation; 3 – initial shaped (cumulative) charge; 4 – flight accelerator; 5 – explosion chamber (underground engine. It may be a serial set of the solid fuel simple rocket disks); 6 - folding hooks; 7 – main shaped (cumulative) chambers (it may be a serial set of the solid fuel shaped explosive semi-spherical disks); 8 – channels for exhaust gas; 9 – tank for a liquid explosives; 10 – 12 – injectors for liquid explosives.

Work. The bomb uses the following method. After delivering the bomb to the vicinity of the target, the accelerator 4 turns on increasing the speed 250 – 500 m/s over a falling speed. The forward cumulative charge produces the narrow channel into the bunker solid protection (it may be armor), injects the liquid explosive into channel and explodes it. The bomb utilizes initially the enormous kinetic energy for initial penetration. After this, the bomb begins to penetrate by itself. Bomb produces the following actions) (fig. 7): (a) Explode the first explosive in main shaped chamber 7 (SFF). Slug creates the canal 1 (fig.7) into concrete or/and soil; (b) Inject the liquid explosive by the very strong jet (strong pressure) into canal; (c) Open hooks 3 and ignite the explosive in the canal. We get the cavity 4 under bomb; (d) remove the exhaust gases from chamber 4 from canals into bomb; (e) remove hooks, explode a first solid fuel disk in chamber 6 of underground rocket engine (or inject a liquid explosive into the engine chamber 5 and ignite). The exhaust rocket gases (explosion) move the bomb into empty cavity 4 and penetrate into concrete (ground); (f)-(g) repeat the actions (a)-(e) while the bomb has shared and rocket disks and the liquid explosive. (h) In final stage (or given depth) the bomb explodes.

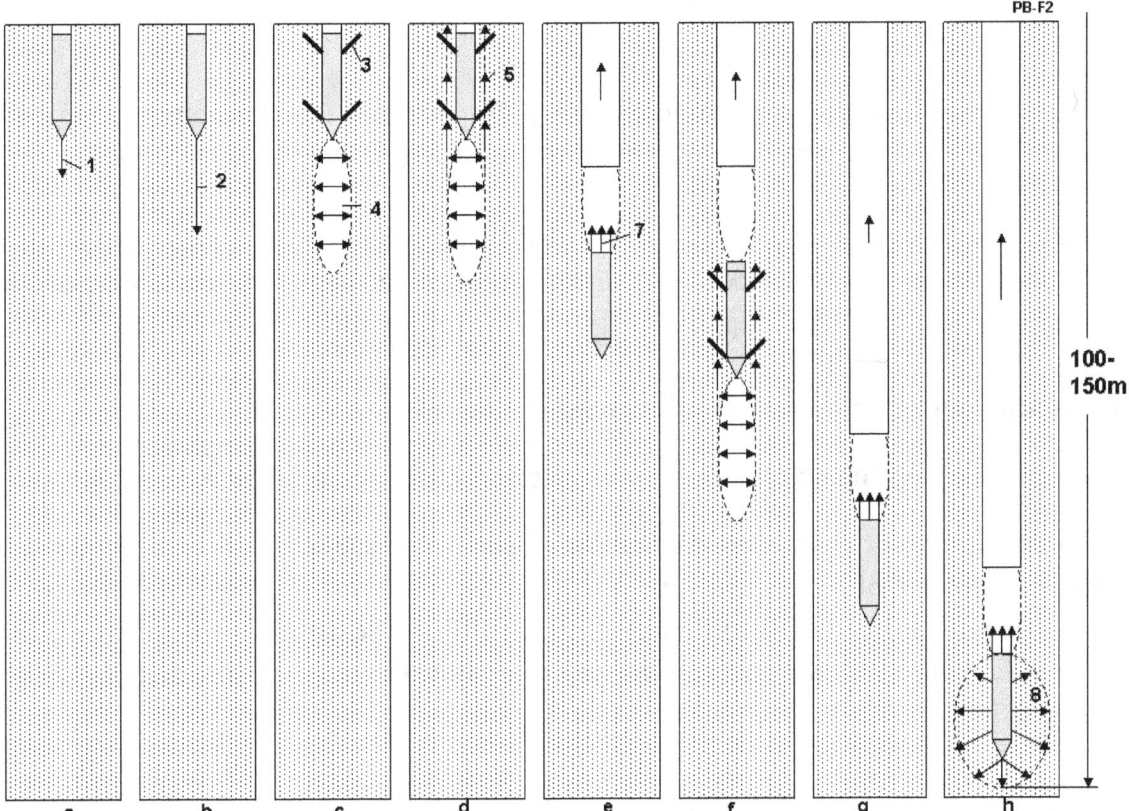

Fig.7. Work of Self-propelled Bomb after the penetration as the conventional penetration bunker bomb into Earth (or protected concrete). Action and notations: (a) Explode the first explosive disk in the main shaped chamber 7 (SFF). Slug creates the canal 1 (fig.2) into concrete; (b) Inject the liquid explosive by the very strong jet (big pressure) into canal; (c) Open hooks 3 and ignite the explosive in the canal. We get the cavity 4 under bomb; (d) delete the exhaust gases from cavity 4 threw canal 8 (fig.6); (e) remove hooks, explode a first solid fuel disk in chamber 6 of the underground rocket engine (or inject a liquid explosive into the engine chamber 5 and ignite). The exhaust rocket gases (explosion) moves the bomb into empty cavity 4 and penetrates into concrete (ground); (f)-(g) repeat the actions (a)-(e) while the bomb has shared and rocket disks and the liquid explosive. (h) In final stage (or given depth) the bomb explodes.

Innovations. Method:

1. Using liquid explosive.
2. Using the initial cumulative charge for destroying the armor cover of object (aim).
3. Multiple using the cumulative (shaped) charges for producing the narrow channels.
4. Injecting the liquid explosive into these channels.
5. Firing (exploding) of this liquid explosive and creating the cavity for bomb (apparatus).
6. Pushing the bomb (apparatus) in given cavity and ground by firing of the charge on the bomb bottom (rocket effect).
7. Repeating this process while there are explosive or while we reach our purpose (bunker/given depth).
8. Exploding bomb.
9. Bomb has forward part which has the cumulative and conventional charges for initial

destruction of a bunker armor protection.

Advantages:
1. Liquid bomb can reach a big additional depth up **70 - 100** m by kinetic energy to the depth received by a current conventional penetration bunker bomb and hundreds additional meters of depth by self-moving .
2. The weight of bomb is about 1.5 – 2.5 tons. That is acceptable for most military aircraft.
3. Method may be used for the super quick drilling of the oil and gas pipe lines. We can add (delivery) the explosive to the apparatus by tube and reach previously unfathomable depths.

Theory, Estimation and Computation of Penetration Bombs

The theory allows estimating the main parameters of the penetration/bunker bomb.

1. Kinetic penetration ability of the bunker bomb. Theory of a penetration the projectile into barrier is very complex. The depth of penetration depends from many values. There are a numerous of methods of computing this but the different methods give different results. That way the best method is testing on dissimilar bunkers. For example, Kinetic penetration ability of the bunker bomb may be estimated by equation:

$$L = \frac{MV^2}{2pS}, \qquad (1)$$

where E is energy, J; L is penetration distance, m; M is mass of bomb, kg; p is average specific drag of medium, N/m²; S is maximal cross section area of bomb, m². For example, if the bomb has mass M = 2000 kg, diameter 0.3 m (S = 0.225 m²) and speed V = 447 m/s, the bomb penetrates L = 80 m into the reinforced concrete having a strong p = 36 MPa (360 atm).

The critical collapsing pressures p for different materials are presented in Table 1.

Table 1. Critical collapsing pressures p for different materials [1].

Material	Density, kg/m³	p, MPa=10 atm	Material	Density kg/m³	p, MPa=10 atm
Reinforced concrete	2000÷2200	4.9 ÷34	Sand	1200÷1600	0.1 ÷ 1
Brick	1600÷1700	7 ÷29	Sandstone	1500 ÷ 1800	1 ÷ 5
Granite	2010 ÷2250	147÷255	Soil, gravel	1500÷2000	1 ÷ 4
Ice	900	1 ÷ 2	Armor (steel)	7900	373 ÷ 412

In War II designers used the following method for calculation the artillery shell penetration in bunker protection.

$$L = 10^{-6} kMV \sin\alpha / d^2, \qquad (2)$$

where L is depth of penetration, m; k is coefficient of penetration from Table 2, M is mass of shell; V is speed of shell at bunker, m/s; d is caliber of gun, m; α is angle between axis of shell and a bunker surface.

Table 2. Coefficient of penetration of artillery shell into the bunker protection [2]

No	Material	k	No	Material	
1	Reinforced concrete	0.7÷1.3	9	Sand	4.5
2	Granite rock without cracking	1.6	10	Clay loam, dry	5
3	Gravel without cracking	2	11	Clay loam, moist	6
4	Stone in cement mortar	2	12	undisturbed soil, the earth's array	6.5
5	Brick-paving stone, dry	2.5	13	Compact clay	7
6	Brick in cement mortar	2.5	14	Bulk sand	9
7	Brick, dry	3	15	Wet clay, wet soil, swamp	10
8	Pine in logs	6	16	filled up the land	13

2. Bomb speed from altitude falling without air drag is
$$V = \sqrt{2gH}, \qquad (3)$$
where $g = 9.81$ m/s² is the Earth acceleration; H is altitude, m. Example, if the bomb fall from altitude $H = 10,000$ m, one gets a speed about $V = 450$ m/s.

3. Maximal bomb fall speed with air drag approximately equals
$$V_m \approx \left(\frac{2gM}{C_d \rho S}\right)^{0.5}, \qquad (4)$$
where C_d is average drag coefficient ($C_d = 0.12 \div 0.3$); $\rho = 1.225$ kg/m³ is air density.

Typical value is approximately $V_m \approx 1400$ m/s. That does not limit the vertical speed of bomb having the good aerodynamic form.

The wing bomb having good ration K (lift force/drag) can convert the part of a horizontal aircraft speed in an additional vertical bomb speed. This part equals
$$V_v \approx \left(V_a^2 - \frac{\pi g r_b}{K}\right)^{0.5}, \qquad (5)$$
where V_v is the horizontal aircraft speed converted in a vertical bomb speed, m/s; V_a is the horizontal aircraft speed, m/s; $g = 9.81$ m/s² is Earth's acceleration; r_b is radius of bomb trajectory from wing, m. For $V_a = 200$ m/s, $r_b = 1000$ m, $K = 10$ the additional (to an altitude bomb speed V (Eq.(3)) speed $V_v = 192$ m/s.

4. If the bomb has wings, the maximal gliding range is
$$R = KH, \qquad (6)$$
where K is ratio lift force to air drag, $K \approx 5 \div 12$. From altitude $H = 10$ km the wing bomb can glide up 120 km with aircraft speed.

5. Additional bomb speed from rocket accelerator is

$$\Delta V = V_g \ln \frac{M_0}{M_f}, \qquad (7)$$

where V_g is speed of rocket exhaust gas, m/s. For a solid fuel rocker one is about V_g = 2300 ÷ 2800 m/s, for a liquid rocket engine V_g = 3100 ÷ 3300 m/s; M_0 is initial rocket mass; M_f is a final rocket mass. Example: if solid fuel rocket spend 1% its mass, one receives speed about $\Delta V \approx$ 25 m/s.

6. The conic shape (cumulative) explosive we can penetrate the good armor

$$b = l(\gamma_j / \gamma_c)^{0.5}, \quad \text{were} \quad l = d / 2\sin\alpha. \qquad (8)$$

Here l is length of shape jet (EFP), m; α is angle between axis and conic cover; γ is density of conic cover and media respectively; d is diameter of shape charge. For α =15° and γ_j / γ_c = 4, value $b \approx 4d$ for strong armor (special steel). Special forms increase l in two times.

For special semi-sphere shape explosive (SFF) the speed of slug can reach tens of kilometers/sec and the small projectile (into shape jet) can reach some kilometers/sec and the length of penetration (canal) in some hundreds d. The length of canal may be estimated by Equation (1) for speed more 2000 m/s.

7. Liner mass m [kg/m] of explosive is needed for increasing canal/cavity up to the radius of bomb r may be calculated by equation:

$$E = mw = \pi r^2 p, \quad m = \frac{\pi p r^2}{w}, \qquad (9)$$

where w is the specific energy of explosive, J/kg; typically w = 4.5 ÷ 6 MJ/kg.

Project

Let us take one configuration of the new bomb with the mass of M = 2000 kg, diameter d = 0.3 m and length 7 m. Bomb has a solid fuel rocket accelerator having mass 5% from the bomb mass (Ma = 100kg). If the bomb drops out from altitude H = 10 km, one gets the additional (to aircraft 220 m/s) speed from falling 447 m/s. The rocket accelerator adds 132 m/s. If total speed is 447+132=579 m/s (without aircraft speed), for reinforced concrete 5000 psi (36 MPa) the initial bomb depth is 80 m.

After the initial kinetic penetration the offered bomb begin a self-penetration actuated by multiple detonations of shaped charges (SFF). The shaped charge penetrates into the soil, and produces a narrow channel with a diameter of 1 cm and a length of 5 - 80 meters (depending on the hardness of the soil: from concrete to sand, 2 - 36 MPa). In this channel the bomb injects a liquid explosive in amounts of from 13 to 234 gr/m, which upon detonation the channel expands to a diameter of 0.3 m. Then the bomb blasts the a charge - up to (1-2)% from the bomb mass at the bottom of the bomb and the bomb get a speed 25 - 50 m/s which pushes it into the cavity – canal after explosion. In the result, the bomb spent only 1-2% of their mass moving into the additional depths of the Earth in 5 - 80 m. This procedure can be repeated by the bomb many times. Bomb can reach depths in the hundreds of meters. When the bomb reaches at a predetermined specific depth or an enemy bunker, it explodes.

Summary

The authors offered the new penetration bomb/projectile (Self-propelled underground bomb) which can move underground in hundreds of meters. This bomb can reach the deepest bunker in the

World. Same design may be used for the self-moving underground apparatus for super quick oil/gas drilling. The reader may find additional relevant information in [3]-[5].

References

[1] Кошкин Н.И., Ширкевич М.Г., Справочник по элементарной физике. Москва, Наука, 1982, стр.44. (Directory of Physic).

[2] Armor http://btvt.narod.ru/4/armor_penetration.htm (in Russian).

[3] Bolonkin A.A., "New Concepts, Ideas, Innovations in Aerospace, Technology and the Human Sciences", NOVA, 2006, 510 pgs. http://www.scribd.com/doc/24057071 , http://www.archive.org/details/NewConceptsIfeasAndInnovationsInAerospaceTechnologyAndHumanSciences

[4] Bolonkin A.A., Femtotechnologies and Revolutionary Projects. Scribd, USA, 2011. 538 p. 16 Mb. http://www.scribd.com/doc/75519828/ http://www.archive.org/details/FemtotechnologiesAndRevolutionaryProjects

[5] Wikipedia, Some background material in this article is gathered from Wikipedia under the Creative Commons license. http://wikipedia.org .

Chapter 16*.
Design of Optimal Regulator
Summary

Current research suggests the use of a liner quadratic performance index for optimal control of regulators in various applications. Some examples include correcting the trajectory of rocket and air vehicles, vibration suppression of flexible structures, and airplane stability. In all these cases, the focus is in suppressing/decreasing system deviations rapidly. However, if one compares the Linear Quadratic Regulator (LQR) solution with optimal solutions (minimum time), it is seen that the LQR solution is less than optimal in some cases indeed (3-6) times that obtained using a minimum time solution. Moreover, the LQR solution is sometimes unacceptable in practice due to the fact that values of control extend beyond admissible limits and thus the designer must choose coefficients in the linear quadratic form, which are unknown.

The authors suggest methods which allow finding a quasi-optimal LQR solution with bounded control which is closed to the minimum time solution. They also remand the process of the minimum time decision.

Keywords: Optimal regulator, minimum time controller, Linear Quadratic Regulator (LQR).

---

*This chapter is declared a work of the U.S. Government and not subject to copyright protection in the USA.

The chapter is accepted as paper AIAA-2003-6638 by 2nd AIAA "Unmanned Unlimited" Systems, Technologies, and Operations- Aerospace, Land, and See Conference and Workshop & Exhibit, San Diego, California, USA, 15-18 Sep 2003.

Introduction

The LQR solution is easily and conveniently written using the Riccati equation as an optimal solution. The scientist who accepts this may be acting as an intoxicated man in a Russian anecdote: one night a man is observed creeping around a streetlight. A passerby asks him, what are you doing? – I lost money. Where did you lose the money? –There at the other end of the street. Then, why are you looking here? – This is where the light is!

The minimum time solution is more complex, however, it can be conveniently determined in many problems by the availability, generally, of high-speed computers. Also, this approaches us with a true minimum time solution.

For an *n*-dimensional problem with one control this solution found in general form in reference [1]. For the two-dimensional case this solution can be presented graphically, see ref. [1]. Methods for other general optimal solutions are offered in [2]-[4].

The LQR solution has three main issues:

1) The selection of the matrix coefficients in the performance index are designer selected and the solution is dependent upon the value of these coefficients.
2) The range of control values can be large in number and this not admissible for practice.
3) The "optimal" LQR solution can be up to 3-6 times worse, then the minimum time solution (see the example in this paper).

If a researcher chooses to use the LQR solution, the authors suggest a method for limiting maximum control (see point 2) as well as for the choice of selecting the coefficients in the performance index. This allows up to a 2-3 times improvement in the performance index (see accompanying examples) and thus makes the LQR solution acceptable in practical applications.

The traditional approach used in the design of a controlled structural system is to design the structure first by satisfying given requirements and then to design the control system. The structure is designed with such constraints placed on weight, allowable stresses, displacements, buckling, general instability, frequency distributions, etc. When the selection of the geometry, cross-sectional area of the members, and material are determined for a specified structure, then the structural frequencies and vibration modes become important input in the design of the control system. Some investigators have written papers discussing an integrated design approach for optimal control. In most references, the control design procedures used, do not take into consideration the limitations on the control forces developed by the actuators, and have not been treated as constraints or design variables. In this paper the problems associated with the selection of the performance index, parameters, weight coefficient in the LQR problem, and limitation of control forces are addressed.

In the following sections, theories for the synthesis of an optimal control laws with a quadratic performance index and bounded control forces are given. This is followed by a SISO (Single Input, Single Output) control problem designed using both approaches for comparison of the end state trajectories, with different bounds placed on control forces. Next, the control system for an idealized wing-box is used to illustrate a design application of the method. A discussion on the application of a control system with bounded control for an integrated design of a structure and control system can be found in ref. [5]. Related articles are [6]-[10].

1. Optimal Control

The general optimal control problem can be described by the following equations

$$I = F_0(x_1, x_2) + \int_{t_1}^{t_2} f_0(t, x, v) dt, \quad dx/dt = f(t,x,v), \quad x(t_1) = x_1, \quad x(t_2) = x_2 \quad (1\text{-}1)$$

where I is the functional (objective function), t is time, x is a n-dimensional vector of state, and v is a p-dimensional vector of control forces. The vector $v \in V$ where V can be a bounded domain. Boundary conditions t_1, t_2, x_1, x_2 are usually given, $(t_1,t_2) \in T$.

The control parameter, v is calculated so that $I = min$. To find the solution to this problem by Method of Deformation of Chapter 2 (α – function), assume the function

$$\psi = \psi(t,x) \quad (1\text{-}2)$$

and write the new function

$$J = A_1 + \int B_1 dt \quad t \subset [t_1, t_2], \quad (1\text{-}3)$$

where

$$A_1 = F_0 + \psi(t_2) - \psi(t_1), \quad B_1 = f_0 - (\partial \psi/\partial x)f - (\partial \psi/\partial t). \quad (1\text{-}4)$$

Here $(\partial \psi/\partial x)$ is n-dimensional vector of partial derivatives. The global minimum is

$$A_0 = \inf_{x_1, x} A_1(x_1, x_2), \quad B_0 = \inf_{v, x} B_1(t, x, v) \quad for \forall t \in T. \quad (1\text{-}5)$$

Depending on the nature of the functions used for ψ, a different set of algorithms for obtaining the infimum can be developed. For example, if Eq.(1-2) takes the form

$$\psi = \lambda(t)x, \quad (1\text{-}6)$$

where $\lambda(t)$ is an n-dimensional vector, the global minimum functions can be written as,

$$A_0 = \inf_{x_1, x_2} (F_1 + \psi(t_2) - \psi(t_1)), \quad B_0 = \inf_{x, v} [f_0 - \lambda f(t,x,v) - (d\lambda/dt)x] = \inf_{x, v} B. \quad (1\text{-}7)$$

Using $\partial B/\partial x = 0$ and Eq.(1-7) gives

$$d\lambda/dt = -\partial H/\partial x, \quad B = \inf_{v \in V} B, \quad v \in V, \tag{1-8}$$

where $H = \lambda f(t,x,u) - f_o$.

Eq. (1-8) can be integrated to find ψ, to obtain the optimal control v and the optimal trajectory $x(t)$. Another way is to enforce the condition

$$B = \inf_{v \in V}\left[f_0 - \frac{\partial \psi}{\partial x}f - \frac{\partial \psi}{\partial t}\right] = 0, \tag{1-9}$$

everywhere in the admissible domain for x. In this case, the equation for particular derivatives can be solved and the syntheses of the optimal control $v = v(t,x)$ and the field of the optimal trajectory in the admissible domain is obtained.

The two control design approaches with constraints on the maximum control forces are discussed in this section. In the first section an objective function for establishing of the minimum time to suppress vibration is discussed and in the second, the quadratic function is minimized.

A. Minimum Time

Since the main purpose of the controller is to suppress vibrations in minimum time, the time for the system to come to rest is taken as the objective function. A functional expression for this can be written

$$I = \int_0^T dt, \quad T = \min \tag{1-10}$$

subject to

$$dx/dt = Ax + bf, \quad x(0) = x_o, \quad x(T) = 0 \tag{1-11}$$

with control force limits

$$|f_i| \le F_i, \quad i = 1,2,\ldots,p. \tag{1-12}$$

This problem can be written in short form as

$$\min I = \int_0^T dt, \quad dx/dt = Ax + bf, \; x(0) = x_o, \; x(T) = 0, \; |f| \le F, \tag{1-13}$$

where x is the state vector of dimension $2n$. A is the $2n \times 2n$ plant matrix, B is $2n \times p$ control matrix, f is the control force vector of dimension p, $x(0)$ is the initial state vector, and $x(T) = 0$ is the final state of the system. B_o, in Eq.(1-7), for this problem can be written as

$$B_o = 1 - \Sigma(\partial \psi/\partial x_i)(dx_i/dt) - (\partial \psi/\partial t) \quad (i = 1,\ldots,n). \tag{1-14}$$

Substituting

$$\psi = \Sigma \lambda_i(t)x_i \quad (i = 1,2,\ldots,2n) \tag{1-15}$$

and Eq.(1-11) into Eq. (1-14) gives

$$B = 1 - \sum_{J=1}^{2n} \lambda_J \left(\sum_{i=1}^{2n} a_{ij} + \sum_{k=1}^{p} b_{jk} f_k \right) - \sum_{j=1}^{2n} \dot{\lambda}_j x_j \quad . \tag{1-16}$$

Taking the partial derivatives of B $(\partial B/\partial x_i)$ gives

$$d\lambda_j/dt = -\sum_j a_{ij}\lambda_j \qquad i = 1,2,...,n\ ,\ j = 1, 2,...,2n\ . \tag{1-17}$$

min B_o gives the control force f

$$f_i = |F_i|\ \text{sign}\ (\Sigma b_{jk}\lambda_k) \qquad i = 1, 2,...,p\ ;\ k = 1, 2,...,2n\ . \tag{1-18}$$

Using Eqs.(1-13),(1-17) and (1-18), the optimal control force $f_i(t)$ and trajectory $x_i(t)$ can be calculated. However the initial $\lambda_i(0)$ for our trajectory with $x(0) = x_o$ is not known. To find $\lambda_i(0)$, any suitable gradient method can be used. For example, if the assume some initial state $\lambda_i(0)$ and integrate Eqs.(1-13),(1-17) and (18), we can calculate the function

$$I = T + \sum_{i=1}^{2n} C_i x_i^2(T), \qquad C_i > 0\ , \qquad (i=1, 2, ...,2n). \tag{1-19}$$

Here C_i are weight coefficients. If $\Sigma C_i x^2_i(T) < C_0$, where C_o is small, the problem can be considered as solved. Time is optimal and $x_i(t)$ is the optimal trajectory which satisfies the final condition $x_i(T) = 0$. If $\Sigma C_i x^2_i(T) > C_o$ we can choose a new $\lambda i(0)$ by any method and repeat the process until it satisfies $\Sigma C_i x^2_i(T) < C_o$.

In practice, a new independent variable τ is introduced as $t = c\tau$, which can be included with Eq.(1-11) to prouder the additional equation

$$dt/d\tau = c\ . \tag{1-20}$$

Additionally, introducing a fixed interval of integration $[0, \tau_1]$ a new set of equations become

$$\min I = \int_0^{\tau} c\,d\tau,\quad dx/dt = (Ax + Bf)c\ .\ x(o) = x_o,\ x(\tau_1) = 0,\ |f| \leq F\ , \tag{1-21}$$

where c is some constant, which is also selected. Eq.(1-19) thus becomes

$$I_1 = \Sigma C_i x^2_i(\tau_1),\quad i = 1,2,...,2n\ . \tag{1-22}$$

For the structural system as defined by Eqs.(1-11)-(1-12) this problem can be solved for the case in which the number of control inputs, p, is equal to the number of modeled structural degrees of freedom, n. However, numerical difficulties would be encountered when this condition is not satisfied. Typical difficulties would be the occurrence of many local minimums, poor convergence, and the need for smaller step sizes.

B. Linear quadratic regulator (LQR) with bounded control

In this case, a performance index, J, is defined as

$$J = \int_0^{\infty} (x^T \underline{Q} x + f^T \underline{R} f) dt \qquad t \in [0,\infty]\ . \tag{1-23}$$

Where $\underline{Q}$ and $\underline{R}$ are state and control weighting matrices. The matrix $\underline{Q}$ must be positive semi definite ($x^T Q x \geq 0$), and $\underline{R}$ must be positive definite ($f^T R f > 0$). The dimensions of $\underline{Q}$ and $\underline{R}$ depend on the size of vectors the x and f, respectively. The matrices $\underline{Q}$ and $\underline{R}$ can be written as

$$\underline{Q} = \sigma Q \tag{1-24}$$

and

$$\underline{R} = (1/\gamma)R^{-1} \tag{1-25}$$

where σ and γ are the design positive variables and Q and R^{-1} are constant identity matrices.

The weighting matrix R is defined in terms of the inverse of the constant matrix R in order to maintain positive definiteness. The function B, Eq. (1-9) for the performance index defined in Eq.(1-23) and the constraint equation Eq.(1-11), become

$$\overline{B} = \inf_f \left[(\sigma x^T Q x + \gamma f^T R f) - \frac{\partial \psi}{\partial x}(Ax + Bf) - \frac{\partial \psi}{\partial t} \right] = 0 \tag{1-26}$$

If V represents an open domain, the function ψ, can be written in the form

$$\psi = x^T P x, \tag{1-27}$$

where P is a $2n$-dimensional unknown matrix.

Substituting ψ Eq. (1-27) into Eq, (1-26), we obtain the equation

$$\sigma Q + PA + A^T P - \gamma PBRB^T P = 0 . \tag{1-28}$$

Equation (1-28) is the Riccati equation. A solution of this equation gives the matrix P and one can find the optimal control force as

$$f = -Gx \tag{1-29}$$

where

$$G = \gamma RB^T P \tag{1-30}$$

Integrating Eq. (1-11) using Eq. (29) to obtain the optimal trajectory for the LQR functional. Eq. (1-29) may give unrealistic values of control depending on the selection of γ. The magnitude of control can be decreased by increasing γ, however, this may cause other perturbations of the system (such as the time it takes the oscillation to decay) to deteriorate.

In order to obtain more realistic results, bounds can be placed on the control force. This can be written as

$$|f_i| \leq F_i, \quad F_i = \text{const}, \quad i = 1, 2, \ldots, p \tag{1-31}$$

where F_i is the magnitude bounding each controller. To obtain an optimal solution, the following restrictions must be satisfied: (1) among these optimal synthesis of the control must exist in the domain of interest, (2) the function B Eq. (1-30) must be convex, and (3) the limits of F may be constant or dependent on time only and F must not be equal to zero at any time (Note: if F is very small a loss in stability can occur). For a solution, the system of Eqs. (1-11) and (1-29) must be integrated along with limits imposed by equation (1-31).

The norm for the displacements or total deviation can be defined by

$$R_x(t) = S = \left[\sum_{i=1}^{n} x_i^2(t) \right]^{1/2} \quad i = 1, 2, \ldots, n \tag{1-32}$$

This norm is zero at the time the deviation is zero, and the structure stops vibrating. In the LQR solution domain this time equals infinity. For studying the behavior and comparison of different control systems, a measure of performances has been used based upon. The time required to reduce the norm of the displacements to 2% of their initial value.

Numerical Examples.

Example 1. SISO problem.

For comparison of systems with different objective functions, a vibrating structure with a single physical degree of freedom was been investigated. This system is described by equation the following set of

$$dx_1/dt = x_2, \quad dx_2/dt = -\omega^2 x_1 - 2\zeta\omega x_2 + cf, \quad x_1(0) = 0, \quad x_2(0) = 1, \quad |f| \leq 1 \quad (1\text{-}33)$$

where $\omega = 2$ is the frequency, $\zeta = 0.03$ is the damping, $c = 1$, and $|f| \leq 1$ is the control.
The problem is solved having an objective function for minimum time as

$$\min T = \int_0^T dt, \quad x_1(T) = 0, \quad x_2(T) = 0 \quad (1\text{-}34)$$

Eqs. (1-17) and (1-18) for the system defined in Eq. (1-33) become

$$d\lambda_1/dt = -\omega^2 \lambda_2, \quad d\lambda_2/dt = \lambda_1 - 2\zeta\omega\lambda_2, \quad f = |F| \operatorname{sign} \lambda_2. \quad (1\text{-}35)$$

Eqs. (1-33)-(1-35) are integrated and the initial values $\lambda_1(0)$, $\lambda_2(0)$ are chosen such that the conditions $x_1(T) = x_2(T) = 0$ are satisfied. The details of the solution scheme are not given here because of space limitations.

The performance for the linear quadratic regulator (LQR) is

$$J = \int_0^\infty \frac{1}{2}\left[\delta_1 x_1^2 + \delta_2 x_2^2\right] dt. \quad (1\text{-}36)$$

Using this performance index and solving the Riccati Eq. (1-29) gives

$$f = 2(c/\gamma)(c_{12}x_1 + c_2 x_2), \quad (1\text{-}37)$$

where

$$c_{12} = -[\omega^2 + (\omega^4 + c_o\delta_1)^{0.5}]/2c_o, \quad c_2 = \{-\zeta\omega + [\zeta^2\omega^2 + (0.25\delta_2 + c_{12})c_o]^{0.5}\}/c_o, \quad c_o = c^2/\gamma.$$

In the case of $\delta = \delta_1 = \delta_2$, the time history depends only on δ/γ. The total deviation is

$$R_x = S = (x_1^2 + x_2^2)^{1/2} \quad . \quad (1\text{-}38)$$

Eq. (1-33) is integrated with control given in Eq. (1-37).

The results of this investigation for the case $T = \min$, $\delta/\gamma = 0.25$ and 100 and no control (open-loop system) are shown in Fig's 1, 2, & 3.

Fig. 1 shows the time history of deviation of x_2. As can be seen, an LQR with $\delta/\gamma = 100$ gives better results ($t = 4$ sec) than an LQR with $\delta/\gamma = 0.25$ (time is more than 15 sec) however an even better result is obtained with an objective function of minimum time. In the last case, oscillations are terminated in 1.5 sec.
Fig. 2 shows the variation of a bounded control force $|f| \leq 1$ for the case of $T=\min$, LQR when $\delta/\gamma = 0.25$ and $\delta/\gamma = 100$. The case LQR ($\delta/\gamma = 0.25$) does not use the full control force, the case LQR ($\delta/\gamma = 100$) uses more of the control force, and case $t = \min$ uses the maximum control force all the time.
Fig.3 shows the time history for the total deviation (R_x) with no control, with an objective function for minimum time and with LQR given by control bounds $|f| \leq 1$.
A structural system with any number of degrees of freedom can be transformed into pairs of equations (1-33)(see later Eq.(1-40)-(1-48)) where every pair is independent from the other. If the number of controls equals the number of degrees of freedom the design approach based on minimum time can be used. However, if the number of controls is less than the number of pairs of equations, the solution for the functional $T = \min$

becomes very complex. In this case, the LQR approach is a variable alternative.

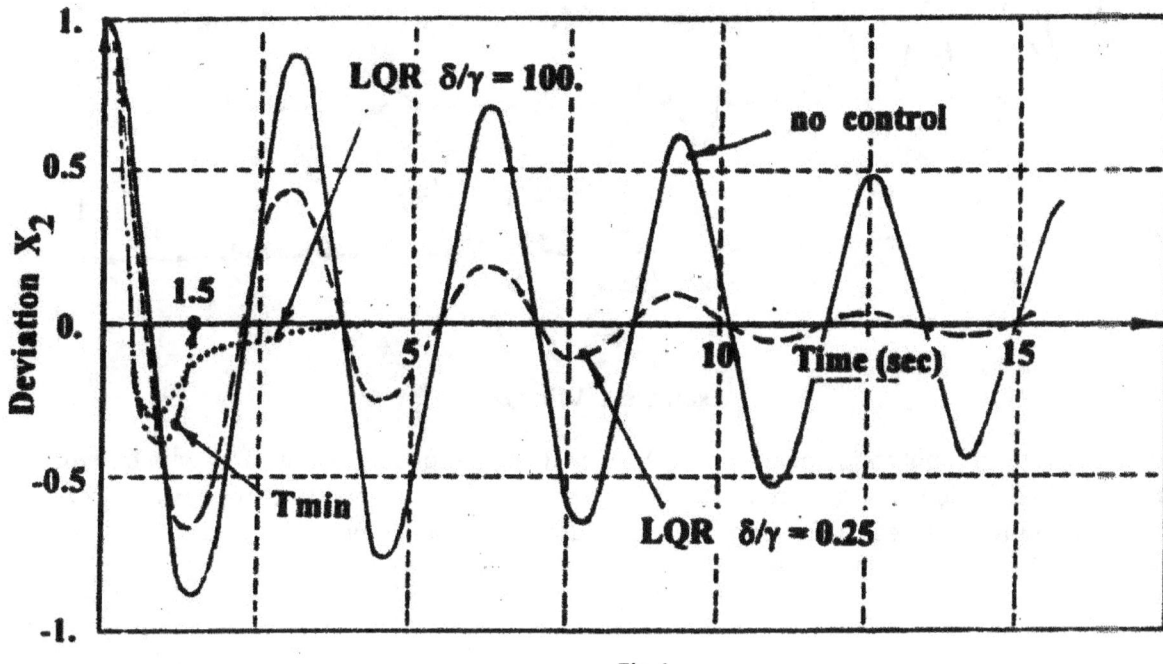

Fig 1

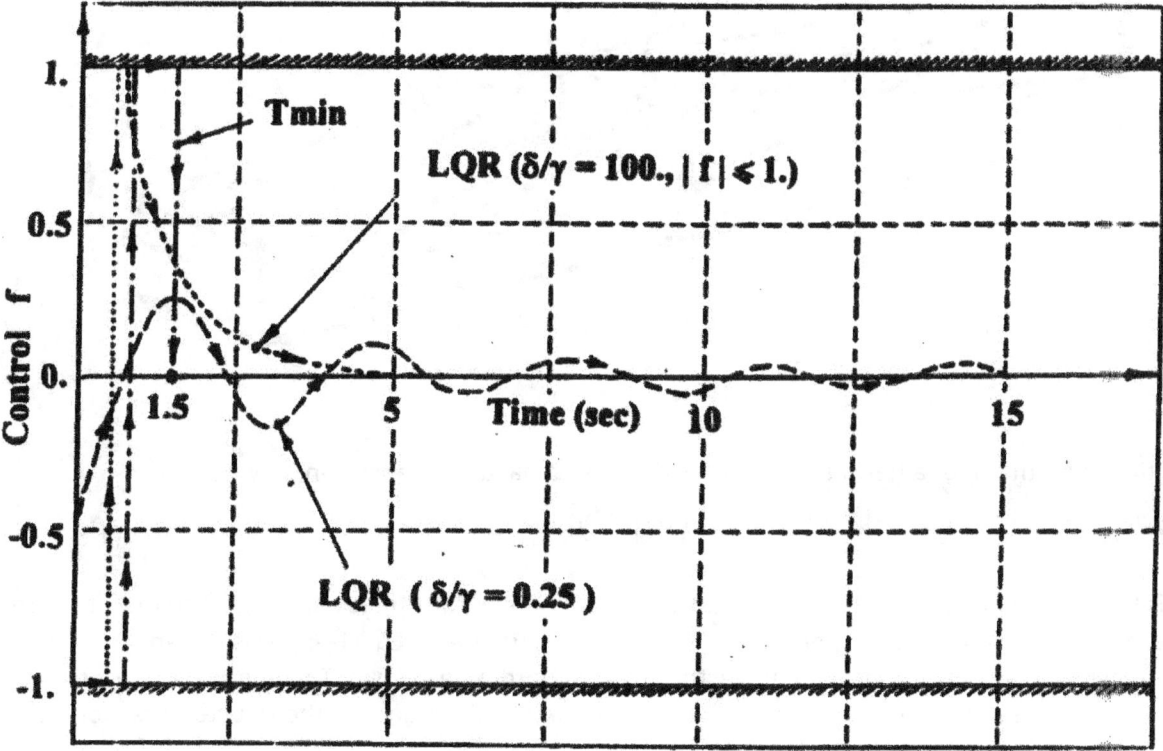

Fig.2

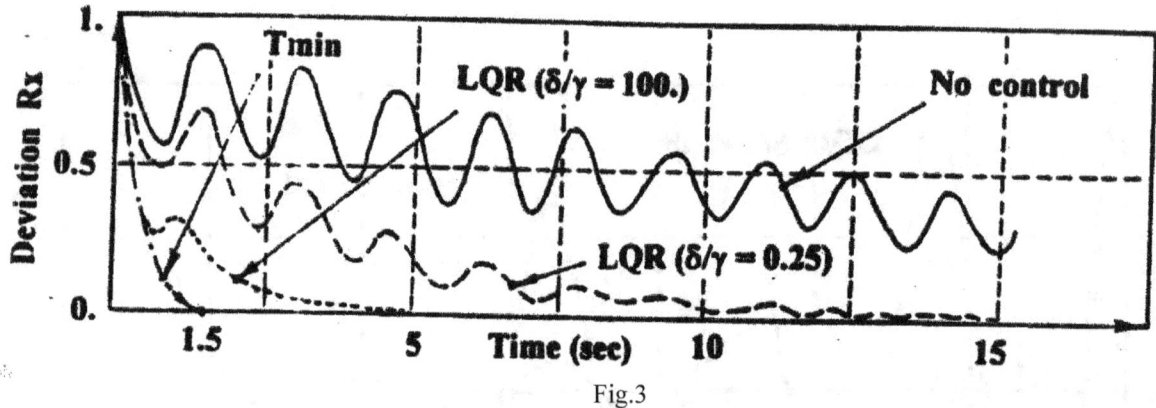

Fig.3

Example 2. Wing Box

In order to illustrate the application of an approach using the linear quadratic regulator with bounded control, the wing box problem in reference [5] is used and shown in Fig. 4. This structure has thirty-two elements and twenty-four degrees of freedom. The structure is a cantilever wing box idealized with bar elements capable of carrying axial loads only.

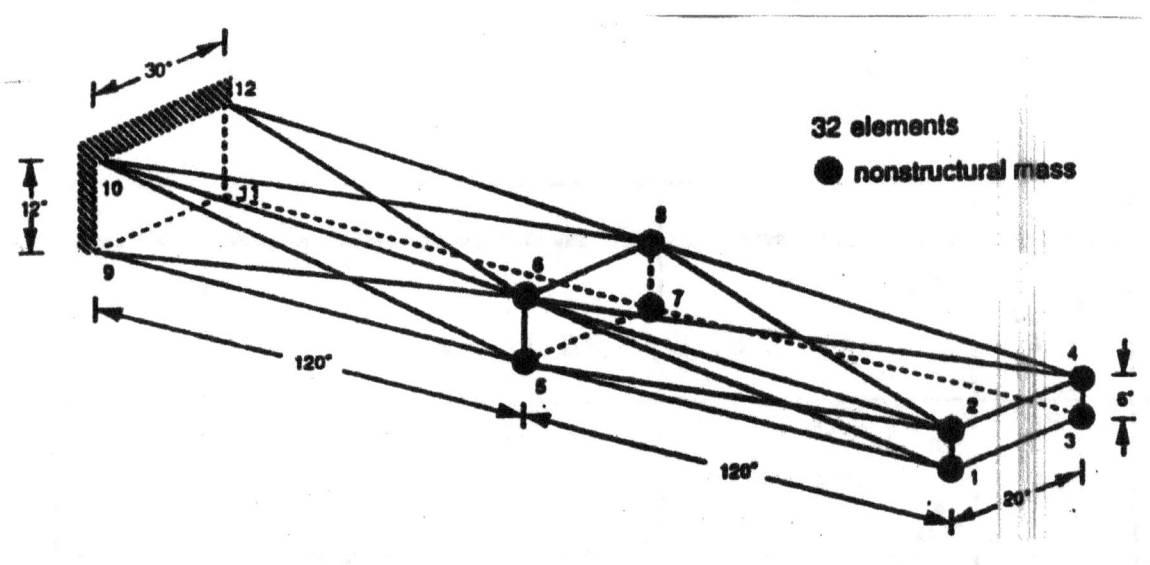

Fig.4.

The equations of motion for a flexible structure with no external disturbance can be written as

$$M\ddot{u} + E\dot{u} + Ku = Df, \qquad (1\text{-}39)$$

where M is the mass matrix, E is the damping matrix, and K is the total stiffness matrix. These matrices are $n_1 \times n_1$, where n_1 is the number of degrees of freedom of the structure. In Eq. (1-39), D is the applied load distribution matrix relating the control input vector f to the coordinate system. The number of elements in vector f is equal to the number of actuators, p. The vector u in Eq.(1-39) defines the structural response.

The coordinate transformation

$$u = [\Phi]\eta \qquad (1\text{-}40)$$

is introduced where η is the modal coordinate system and [Φ] is the $n_1 \times n_1$ modal matrix. Using Eq. (1-40), Eq. (1-39) can be transformed into n_1 uncoupled equations. These can be written as

$$\overline{M}\ddot{\eta} + \overline{E}\dot{\eta} + \overline{K}\eta = [\Phi]^T Df \tag{1-41}$$

where

$$\overline{M} = I = [\Phi]^T M[\Phi]$$
$$\overline{E} = [2\zeta\omega] = [\Phi]^T E[\Phi] \tag{1-42, 1-44}$$
$$\overline{K} = [\omega^2] = [\Phi]^T K[\Phi]$$

The matrices $\overline{M}, \overline{E},$ and $\overline{K}$ are diagonal square matrices, ω is the vector of structural frequencies, and ζ is the vector of modal damping factors. The modal matrix [Φ] is normalized with respect to the mass matrix. The control analysis is performed by reducing the second-order uncoupled equation [Eq.(1-41)] to a first-order equation. Only n of n_1 uncoupled equations are used for the control system design. This can be achieved by using the transformation

$$x_{2n} = \begin{bmatrix} \eta \\ \dot{\eta} \end{bmatrix}_{2n} \tag{1-45}$$

where x is the state variable vector of size $2n$. This gives

$$\dot{x} = Ax + Bf \tag{1-46}$$

where A is a $2n \times 2n$ matrix and B is a $2n \times p$ input matrix. The A matrix and the input matrices are given by

$$A = \begin{bmatrix} 0 & I \\ -\omega^2 & -2\xi\omega \end{bmatrix} \tag{1-47}$$

$$B = \begin{bmatrix} 0 \\ \Phi^T D \end{bmatrix} \tag{1-48}$$

The state output equation is given by

$$y = Cx \tag{1-49}$$

where y is a $q \times 1$ output vector, C is a $q \times 2n$ output matrix, and q is equal to the total number of sensors. If the number of sensors and actuators equal and collocated, then $q = p$ and

$$C = B^T. \tag{1-50}$$

For this structure, Young's modulus and weight density are assumed to be equal to 10.5×10^6 lbs/in^2 and 0.1 lbs/in^3, respectively. The actuators and sensors are assumed embedded in the structural elements and are collocated. The actuators are assumed to apply forces along the axial directions providing both out of plane, in plane and twist control for the structure. It is assumed that all structural modes have 1% structural damping and thus ζ in Eq. (1-9) was 0.01.

The control system utilizes four actuators and sensors collocated in the four members at the tip of the structure connecting nodes 1-2, 3-4, 1-3 and 2-4 respectively. Non-structural masses are located at nodes 1 through 8. Their magnitudes are 0.5 slugs at nodes 1 and 2; 1.5 slugs at nodes 3 and 4; 2.5 slugs at node 5 and 7.0 and 1.0 slugs at nodes 6 and 8 respectively. For the 24 structural degrees of freedom, the full order state space matrix in Eq. (1-11) is 48 x 48. Since there are four actuators and sensors, the input matrix B and output matrix C are 48 x 4 and 4x48, respectively. The cross-sectional areas of the rod elements were equal to 0.1 in^2. The weighting matrices Q and R in Eq. (1-28), (1-29) were equal to the identity matrix.

The four values of the weighting parameter ratios σ/γ selected for this study are 0.1, 1.0, 100 and 1000, respectively. The maximum control forces generated by the four actuators are given in Table 1.

Table 1. Calculated cases

Control bound F	∞	0.5	0.15	0.05
$\delta/\gamma = 0.1$	+	+	+	+
$\delta/\gamma = 1$	+	+	+	+
$\delta/\gamma = 100$	+	+	+	+
$\delta/\gamma = 1000$	+	+	+	+
No control	+			

The initial condition used for designing the controllers is a unit displacement at node 1 in the z-direction. This condition is used for all cases and also to obtain the response curves. The response curves are given for only a few cases because of space limitations. The three limits on the maximum allowable control forces are set equal to 0.5, 0.15, and 0.05 respectively. The different cases considered are summarized in Table 2.

Table 2. Maximum actuator forces

	Actuator #			
Value δ/γ	1	2	3	4
$\delta/\gamma = 0.1$	0.05	0.05	0.07	0.03
$\delta/\gamma = 1.0$	0.20	0.24	0.31	0.12
$\delta/\gamma = 100$	1.31	1.89	3.30	1.23
$\delta/\gamma = 1000$	2.95	2.25	8.25	3.73

In the case of $\sigma/\gamma = 0.1$ the maximum actuator forces are less than 0.15, and for $\sigma/\gamma = 1.0$, they are less than 0.5. Fig. 5 shows the time history of the displacement norm without control bound for the four values of σ/γ and without control.

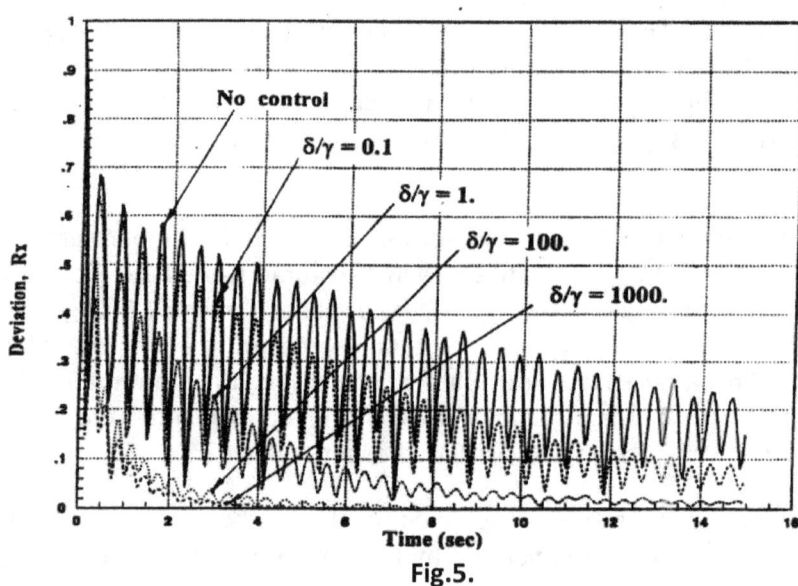

Fig.5.

The maximum value of the displacement norm as a function of time is shown in Fig. 6.
The time required to decrease the displacement norm to 2% of its initial value 1.0 is shown in Fig. 7.

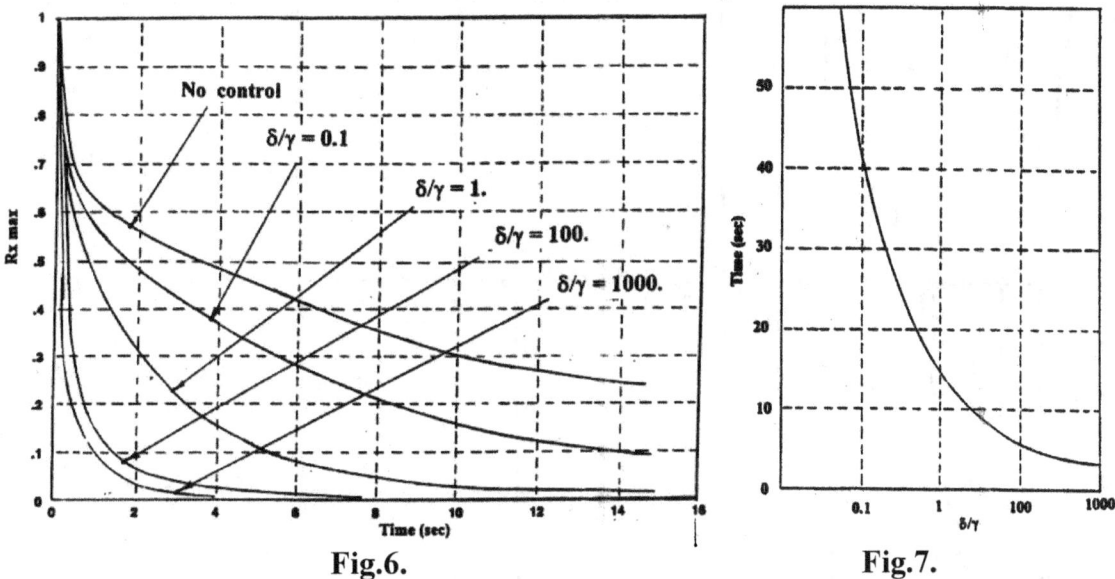

Fig.6.   **Fig.7.**

In the case of no-control, the total time needed to reduce the displacement norm to two percent of the initial value is larger than 100 seconds. The variation in the control force in actuator 1 as a function of time for σ/γ equal to 100 and 1 is shown in Fig. 8.

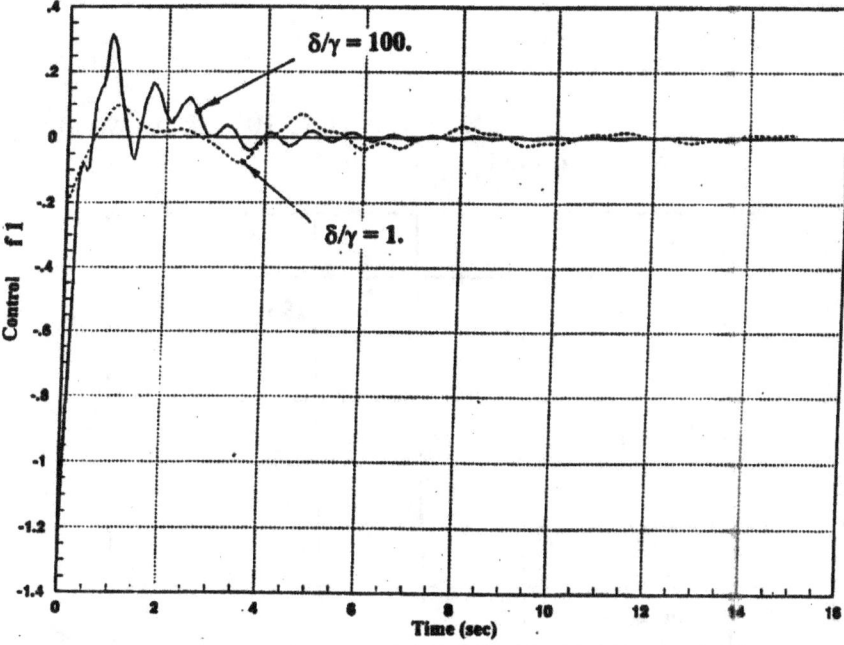

Fig.8. (2.47 Mb)

Fig. 9 shows the time history of control force in actuator 1 with the upper bound equal to 0.15 for σ/γ = 100. The upper bound is enforced on all the actuators.

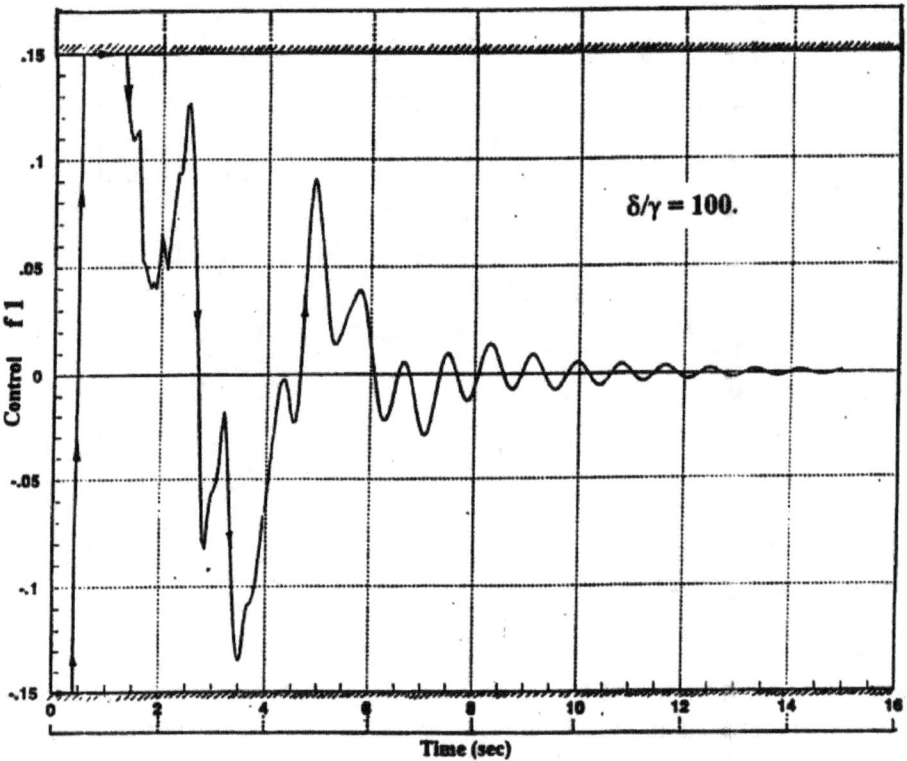

Fig.9.

The changes in the displacement norm with time for δ/γ equal to 100 are shown in Fig. 10 for the case of control bound equal to 0.15 and without bound.

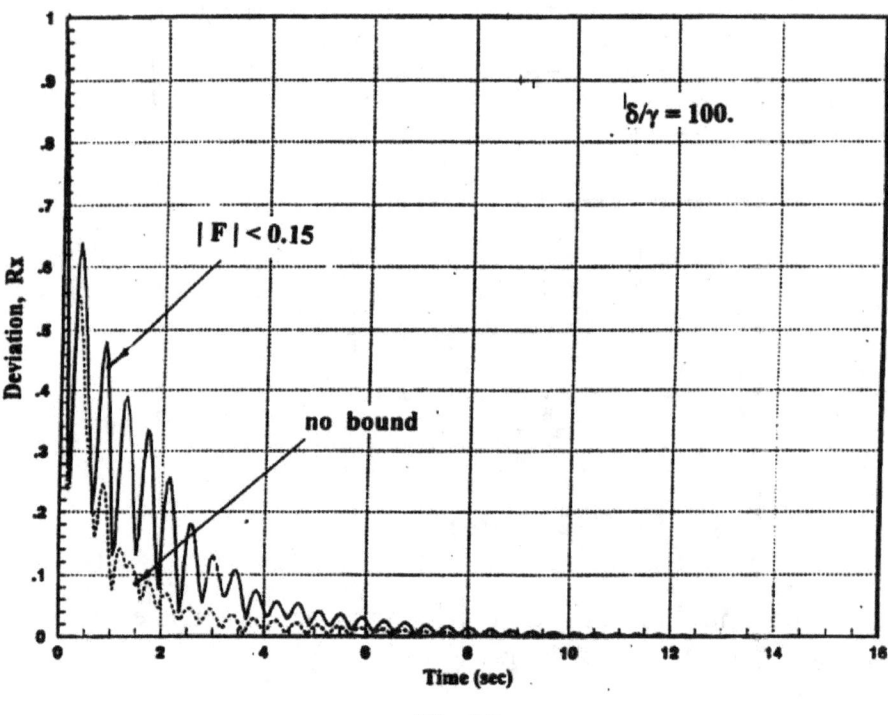

Fig.10.

Fig. 11 shows the total time required to reduce the displacement norm to 0.02 for three values of σ/γ and four values of control bound. As the control bound decreases more time is needed to reduce the displacement norm to 0.02 for a given value of σ/γ. The maximum root mean square response for different cases is shown in Fig. 12.

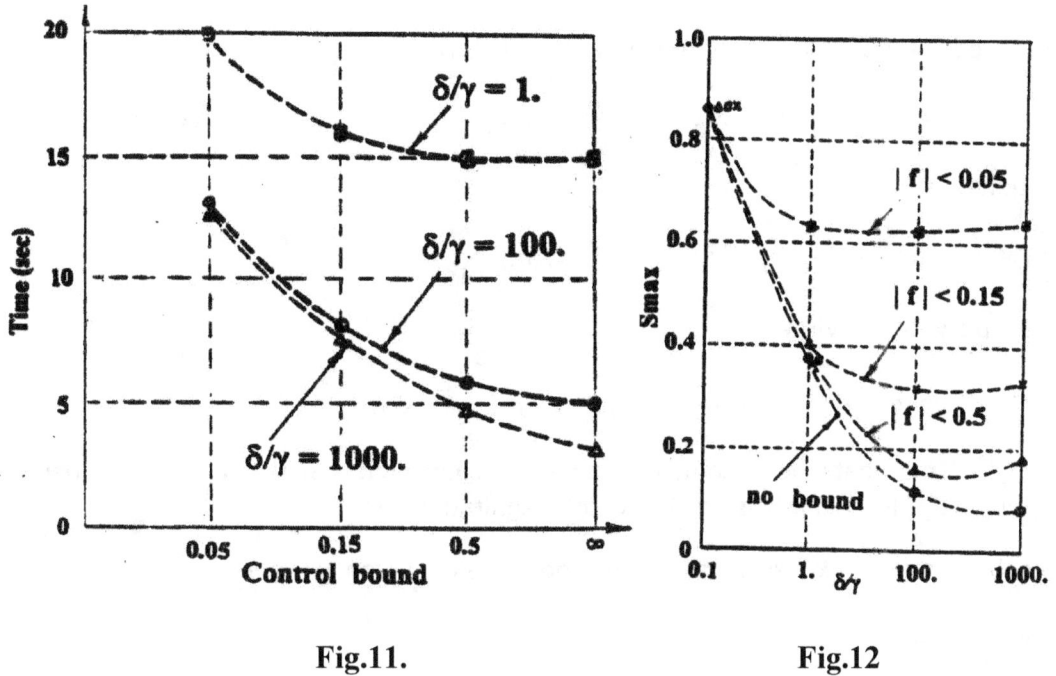

Fig.11.   **Fig.12**

2. Solution of general linear optimal problem for one control

Now consider the general optimal linear regulator problem with an objective function of minimum time and one control parameter.

Problem Statement. The system is described by a linear differential equation in vector form as,

$$\dot{x} = Ax + Lu \qquad (2\text{-}1)$$

where $x = (x_1, x_2, ..., x_n)$ is the *n*-dimensional state vector, $A = \|a_{ij}\|$ a *n*-dimensional square matrix of constant coefficients, *L* a column vector which contains $l_1, l_2, ..., l_n$; *u* a limited control, $|u| \leq \zeta$, $\zeta > 0$; $x(0) = x_o$, $x(t_k) = x_k$ the initial and final condition, $T = t_k$ represent the end time of process, $t_o = 0$.

It is known the control can have only boundary value in linear system and, if eigenvalues of matrix *A* is real numbers, the system has only maximum *n-1* switches [7].

Problem solution. The characteristic equation is $|A - \lambda E| = 0$, where *E* is an unit *n*-dimensional matrix, λ is eigenvalues of matrix *A*.

Case A. All eigenvalues λ are real, different, and not equal zero. Using

$$y_i = \sum_{j=1}^{n} e_{ij} x_j \quad (i=1,2,...,n), \quad e_{ij} = const_{ij}$$

can convert the equations (2-1) to canonical form

$$\dot{y}_1 = \lambda_1 y_1 + b_1 u; \quad \dot{y}_2 = \lambda_2 y_2 + b_2 u; \quad \ldots \quad \dot{y}_n = \lambda_n y_n + b_n u; \qquad (2\text{-}2)$$

with boundary conditions $y_i(0) = y_{i0}$; $y_i(t_k) = y_{ik}$.

The optimal control $u = \pm \zeta$ is constant everywhere. If a new variable $z_i = \lambda_i y_i + b_i u$ is introduced, it is possible to write equation (2-2) in form

$$\dot{z}_1 = \lambda_1 z_1; \quad \dot{z}_2 = \lambda_2 z_2; \quad \ldots; \quad \dot{z}_n = \lambda_n z_n; \qquad (2\text{-}3)$$

A solution of equation (2-3) is

$$z_i = \bar{c}_i e^{\lambda_i t} \quad (i = 1, \ldots, n).$$

Returning to the variable y we can write

$$y_i = c_i e^{\lambda_i t} - b_i u / \lambda_i \quad (i = 1, \ldots, n); \quad c_i = \bar{c}_i / \lambda_i; \quad \lambda_i \neq 0. \qquad (2\text{-}4)$$

Consider the value y_1. The moment when a control parameter is changed it is marked an index "i" below and right and left from point t_i by plus and minus sign on top of magnitudes.

Let us suggest, that the control has $k\text{-}1$ switches. From continuous condition we have $y_i^- = y_i^+$. Therefore we have

$$c_{ji}^- e^{\lambda_j t_i} - b_i u_i^- / \lambda_i = c_{ji}^+ e^{\lambda_j t_i} - b_i u_i^+ / \lambda_i \quad (i = 1, \ldots, n). \qquad (2\text{-}5)$$

From (2-5)

$$c_{ji}^+ = c_{ji}^- + e^{-\lambda_j t_i}(u_i^+ - u_i^-) b_j / \lambda_i \quad (i = 1, \ldots, k\text{-}1). \qquad (2\text{-}6)$$

The value $c_{j,i+1}^- = c_{j,i}^+$. From (2-6), we get

$$c_{jk}^- = c_{j0}^+ + \sum_{i=1}^{k-1} \frac{b_j}{\lambda_i} e^{-\lambda_j t_i}(u_i^+ - u_i^-). \qquad (2\text{-}7)$$

From the first equation (2-4) and boundary conditions for y_i, we find

$$c_{jk}^- = e^{-\lambda_j t_k}(y_{jk} + \frac{b_j}{\lambda_j} u_k^-); \qquad c_{j0}^+ = y_{j0} + \frac{b_j}{\lambda_i} u_0, \qquad (2\text{-}8)$$

where $u_0 = u_0^+$.

Substituting (2-8) to (2-7) we obtain,

$$\sum_{i=1}^{k-1} \frac{b_j}{\lambda_j} e^{-\lambda_j t_i}(u_i^+ - u_i^-) - e^{-\lambda_j t_k}(y_{jk} + \frac{b_j}{\lambda_j} u_k^-) = -y_{j0} + \frac{b_j}{\lambda_i} u_0 \quad (j=1,\ldots,n) \qquad (2\text{-}9)$$

These equation (2-9) satisfies for all y_i $i=1,...,n$.

If to divide the right and left parts of equation (2-9) by $(-2b_ju_0/\lambda_i)$, we find,

$$e^{-\lambda_1 t_1} - e^{-\lambda_1 t_2} + ... - (-1)^k e^{-\lambda_1 t_{k-1}} + \frac{1}{2}\left[(-1)^{k-1} + y_{1k}\lambda_1/b_1u_0\right]e^{-\lambda_1 t_k} = \frac{1}{2}\left[1 + y_{10}\lambda_1/b_1u_0\right];$$

$$e^{-\lambda_2 t_1} - e^{-\lambda_2 t_2} + ... - (-1)^k e^{-\lambda_2 t_{k-1}} + \frac{1}{2}\left[(-1)^{k-1} + y_{2k}\lambda_2/b_2u_0\right]e^{-\lambda_2 t_k} = \frac{1}{2}\left[1 + y_{20}\lambda_2/b_2u_0\right];$$

$$\cdots\cdots\cdots\cdots\cdots\cdots\cdots\cdots\cdots\cdots \qquad (2\text{-}10)$$

$$e^{-\lambda_n t_1} - e^{-\lambda_n t_1} + ... - (-1)^k e^{-\lambda_n t_{k-1}} + \frac{1}{2}\left[(-1)^{k-1} + y_{nk}\lambda_n/b_nu_0\right]e^{-\lambda_n t_k} = \frac{1}{2}\left[1 + y_{n0}\lambda_n/b_nu_0\right],$$

where $k=n$.

Noting that $e^{-t_i} = w_i$, equations (2-10) can be written as

$$w_1^{\lambda_1} - w_2^{\lambda_1} + ... - (-1)^k w_{k-1}^{\lambda_1} + \frac{1}{2}\left[(-1)^{k-1} + y_{1k}\lambda_1/b_1u_0\right]w_k^{\lambda_1} = \frac{1}{2}\left[1 + y_{10}\lambda_1/b_1u_0\right],$$

$$w_1^{\lambda_2} - w_2^{\lambda_2} + ... - (-1)^k w_{k-1}^{\lambda_2} + \frac{1}{2}\left[(-1)^{k-1} + y_{2k}\lambda_2/b_2u_0\right]w_k^{\lambda_2} = \frac{1}{2}\left[1 + y_{20}\lambda_2/b_2u_0\right],$$

$$\cdots\cdots\cdots\cdots\cdots\cdots\cdots\cdots \qquad (2\text{-}11)$$

$$w_1^{\lambda_n} - w_2^{\lambda_n} + ... - (-1)^k w_{k-1}^{\lambda_n} + \frac{1}{2}\left[(-1)^{k-1} + y_{nk}\lambda_n/b_nu_0\right]w_2^{\lambda_n} = \frac{1}{2}\left[1 + y_{n0}\lambda_n/b_nu_0\right].$$

Equation (2-11) is solved in order to find $w_i = w_i(y_o)$. Returning to the original variable x, we can write

$$t_i = -\ln w_i(x).$$

where x_o represents the initial point x.

Equations (2-11) are a set of algebraic equations. From boundary conditions, we know that

$$t_k \geq t_{k-1} \geq ... \geq t_1 > 0.$$

That implies that

$$0 < w_k \leq w_{k-1} \leq ... \leq w_1 \leq 1.$$

For control $u = \pm\xi$. This implies that equations (2-11) must be solved twice. If $x(t_k) = 0$ (this means $y(t_k) = 0$), the second solution is symmetric about the origin.

The solution of equation (2-11) is easier to evaluate then the classical optimal control solution. In classical theory a researcher must solve a boundary problem for a set of given differential equations and also find a set of unknown Lagrange multipliers. In using equation (2-10) the researcher first establish the required time increments based upon knowledge of the physical situation.

To find the switch surfaces, for $t_1 = 0$, implies thus the trajectory is located on the first switch surface. In this case in equation (2-11) $e^{-t_1} = w_1 = 1$. We then set about solving the first $n-1$ equations (2-11) for $w_2, w_3, ..., w_k$ and substitute these solutions into the last equation. This leads to an equation $\Phi_1(y) = 0$. By substituting for y we can find $N_1(x)$. This is the first $(n-1)$-dimensional switch surface.

Next by substituting $w_1 = 1$, $w_2 = 1$ in the first $n-1$ equations, and solving the first $n-2$ equations for $w_3, w_4, ..., w_n$ one can obtain solutions and substitute them into last equation. We thus can find a hyper surface $N_2(x) = 0$. The intersection of this hyper surface with $N_1(x)$ creates the second $(n-2)$-dimensional switch surface. Other switch surfaces can be found in as similar way.

Such an optimal control result can be easily found. In selecting u_o, when the state point reaches the switch surface $N_1(x) = 0$, $u_1 = -u_o$. When the state point reaches the switch surface $N_2(x) = 0$, $u_2 = -u_1$ and so on.

If time is deleted from any two of equation (2-4), we obtain a projection of the trajectory on the surface $y_i y_j$

$$y_i = \bar{c}_i (y_i + b_i u / \lambda_i)^{\lambda_i / \lambda_j} - b_i u / \lambda_i, \qquad \bar{c}_i = c_i / c_j^{\lambda_i / \lambda_j}.$$

From (2-2) we can find the boundaries of instability, for positive eigenvalues. For example, if $\lambda_i > 0$, $b_i > 0$, then $y_i(t_k) = 0$. The necessary and sufficient condition unstable solution is given by

$$y_i > \frac{b_i}{\lambda_i} \xi; \qquad y_i < -\frac{b_i}{\lambda_i} \xi.$$

We have only considered cases when the eigenvalues are real, different, and non-equal to zero. Additional cases have been considered in reference [6].

Example. Taking any two of equations (2-1) with eigenvalues λ_1, λ_2 ($\lambda_1 \neq \lambda_2$, $\lambda_1 \neq 0$, $\lambda_2 \neq 0$, $\lambda_1 < \lambda_2$), $x(t_k) = 0$, a canonical form of the equations can be expressed as,

$$\dot{y}_1 = \lambda_1 y_1 + u; \qquad \dot{y}_2 = \lambda_2 y_2 + u; \qquad y(0) = y; \quad y(t_k) = 0; \quad |u| \leq 1. \tag{E1}$$

Equation (2-11) for $u_o = 1$ can be written as

$$w_1^{\lambda_1} - \frac{1}{2} w_2^{\lambda_1} = \frac{1}{2} \lambda_1 y_1 + \frac{1}{2}; \qquad w_1^{\lambda_2} - \frac{1}{2} w_2^{\lambda_2} = \frac{1}{2} \lambda_2 y_2 + \frac{1}{2}. \tag{E2}$$

For $w_1 = 1$ (simplifying for the case $t_1 = 0$), w_2 can be obtained from (E2) and thus

$$y_2 - \frac{1}{\lambda_2} \left[1 - (1 - \lambda_1 y_1)^{\lambda_2 / \lambda_1} \right] = 0. \tag{E3}$$

For $u_o = -1$ equations (2-11) are

$$w_1^{\lambda_1} - \frac{1}{2} w_2^{\lambda_1} = -\frac{1}{2} \lambda_1 y_1 + \frac{1}{2}; \qquad w_1^{\lambda_2} - \frac{1}{2} w_2^{\lambda_2} = -\frac{1}{2} \lambda_2 y_2 + \frac{1}{2}. \tag{E4}$$

Taking $w_1 = 1$, and using w_2 from (E2), we find,

$$y_2 + \frac{1}{\lambda_2} \left[1 - (1 + \lambda_1 y_1)^{\lambda_2 / \lambda_1} \right] = 0. \tag{E5}$$

Using a continuity condition $y_1(t_k) = y_2(t_k)$, the relations (E3), (E5) can be written as one relation

$$y_2 - \frac{sign\, y_1}{\lambda_2}\left[1-(1-\lambda_1 y_1 sign\, y_1)^{\lambda_1/\lambda_2}\right]=0. \tag{E6}$$

If $y_1 = 0$, $y_2 > 0$, then the relation (E6) is greater then zero. From (E1) we see: y_2 will be decrease faster if $u = -1$ for $y_2 > 0$ and $u = +1$ for $y_2 < 0$. This implied that

$$u = -sign\, \lambda_2\, sign\left\{y_2 - \frac{sign\, y_1}{\lambda_2}\left[1-(1-\lambda_1 y_1 sign\, y_1)^{\lambda_1/\lambda_2}\right]\right\}. \tag{E7}$$

To find the equations for optimal trajectories. Referring equations (2-4),(2-11) we find

$$y_1 = c_1 e^{\lambda_1 t} - u/\lambda_1\,;\quad y_2 = c_2 e^{\lambda_2 t} - u/\lambda_2\,;\quad y_2 = c\,(y_1 + u/\lambda_1)^{\lambda_2/\lambda_1} - u/\lambda_2\,. \tag{E8}$$

The last equation in (E8) gives information in the trajectories as shown in figure 13.
These trajectories depend upon the signs of λ_1, λ_2. For then $\lambda_1 > 0$, $\lambda_2 < 0$ the non-stability region is $|y_1| > \zeta\lambda_1$.
For $\lambda_1 < 0$, $\lambda_2 > 0$ the non-stability region is $|y_2| > \zeta\lambda_2$.

In fig.14 also shown optimal trajectories. Once again they depend up on the signs of λ_1, λ_2. Returning to the variables x, the picture 14 is affined deformity.

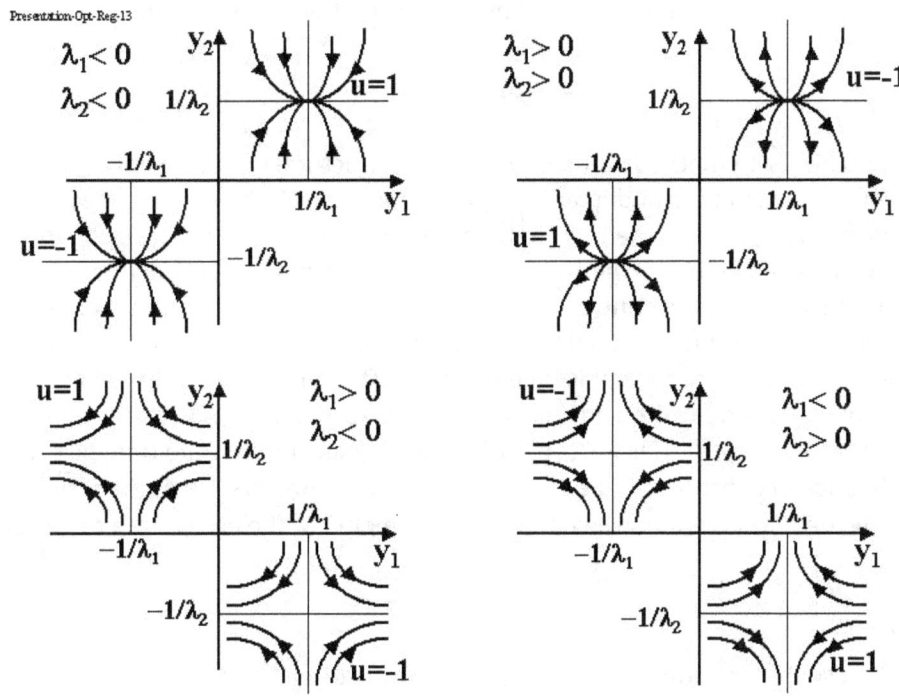

Fig13.

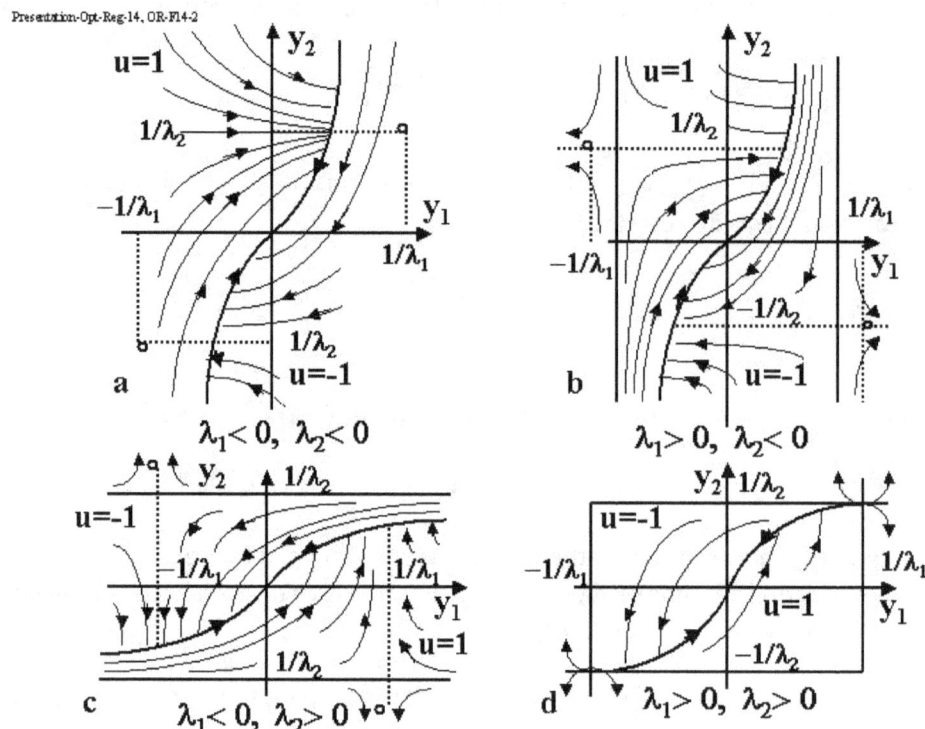

Fig.14.
The offered method allows capture of optimal control.

Summary

Two optimum control design methods for suppression of structural vibration having bounded constrains have been compared. The minimum time and quadratic performance index have been used as objective functions. The second approach leads to use of the LQR methodology with bounded control. The introduction of a minimum time controller can be used when the number of actuators equals the number of structural degrees of freedom used in the design of the control system. When the number of actuators is less than the number of degrees of freedom, the minimum time controller becomes mathematically complicated and has been found to be difficult to solve due to the presence of local minimums. The minimum quadratic function controller, with bounded control, can be designed with a fewer number of actuators. A SISO structural control design problem has been solved using both approaches for comparison of trajectories and the time needed to suppress vibrations. The influence of control limitations and the weight coefficient σ/γ of the structure have been studied. Results indicate that an optimal selection of the weight coefficient σ/γ can decrease the suppression time up to 2-4 times.

Recommendations

If possible, the researcher should try to design the controller for minimum time. If it is very difficult, he can design LQR controller. However in this case the researcher must:

1. Consider limits on the maximum value of the control force.
2. Find the optimal ratio σ/γ of the weight coefficients.
3. Solve (numerically) at least one time the real (minimum time) problem and compare what may be luck is loss from changing the T_{min} problem to the LQR problem.

References

1. A.A. Bolonkin, Solution general linear optimal problem with one control. Journal "Pricladnaya Mechanica", v.4, #4, 1968, pgs. 111-122. Moscow (in Russian).
2. A.A. Bolonkin, New methods of optimization and their application, Moscow, Technical University named Bauman, 1972, pgs.220 (in Russian).
3. A.A. Bolonkin, "Special Extrema in Optimal Control Problems", Akademiya Nauk, Izvestia, Theknicheskaya Kibernetika, No.2, March-April, 1969, pp.187-198. See also English translation in Eng.Cybernetics, n.2, March-April,1969, pp.170-183.
4. A.A. Bolonkin, "A New Approach to Finding a Global Optimum", "New Americans Collected Scientific Reports",Vol.1, 1991. the Bnai Zion Sa Scientists Division, New York.
5. A.A. Bolonkin, N.S. Khot, Optimal Structural Control Design, IAF-94-1.4.206, 45th Congress of the International Astronautical Federation, World Space Congress-1994. October 9-14,1994/Jerusalem, Israel.
6. V. Boltyanski, A. Poznyak, Linear multi-model time-optimization, Journal "Optimal Control Applications and Methods",Vol.23, Issue 3, 2002, pp.141-161.
7. Yunying Mao, Zeyi Liu, The optimal feedback control of the linear-quadratic control problem with a control inequality constraint, Journal "Optimal Control Applications and Methods",Vol.22, Issue 2, 2001, pp.95-109.
8. Heping Hua, Numerical solution of optimal control problems, Journal "Optimal Control Applications and Methods",Vol.21, Issue 5, 2000, pp.233-241.
9. H.Singh, R.H. Broun, D.S. Naidu, Unified approach to linear quadratic regulator with time-scale property, Journal "Optimal Control Applications and Methods",Vol.22, Issue 1, 2001, pp 1-16..
10. B.J. Driessen, N. Sadegh, Minimum-time control of systems with Coulomb friction: near global via mixed integer linear programming, Journal "Optimal Control Applications and Methods",Vol.22, Issue 2, 2001, pp.51-62.

Nomenclature

A is the $2n \times 2n$ plant matrix in liner problem
a_{ij} is members of matrix A
B is $2n \times p$ control matrix in liner problem
b_{jk} is members of matrix B
C is $q \times 2n$ out matrix
C_i are weight coefficients
c is constant
F_i is the magnitude of the bounds for each controller
F_0 is function of initial conditions
f is the control force vector of dimension p
$|f| \leq 1$ is the control in linear problem
H is Hamiltonian
I is the functional (objective function),
$\overline{M}, \overline{E}, \overline{K}$ are diagonal square matrices
P is a $2n$-dimensional unknown matrix
Q is state weighting matrices
R is control weighting matrices
$R_x(t)$ is norm of displacement
T is final time
t is time (variable)
$t_1, t_2,$ - boundary condition

u is the vector defines the structural response.
v is a p-dimensional vector of control forces
x is a n-dimensional vector of state in general problem,
x is the state vector of dimension $2n$ in linear problem.
$x(0)$ is the initial state vector
$x(T)$ is the final state of the system.
x_1, x_2 - boundary condition
ζ is the vector of modal damping factors
$\lambda(t)$ is a n-dimensional vector unknown coefficient
λ is eigenvalues of matrix A
$\psi = \psi(t,x)$ is special function
ω is the vector of structural frequencies

Chapter 17
Impulse solutions in optimal problems

Abstract

The author considers the optimization problem named 'the impulse regime', when the control can have for a short time an instantaneous infinity value and the phase variables have gaps. In mathematics these mean: the variables are not continuous, not differentiable. The variable calculation and Pontryagin principle are not applicable. These problems are in space trajectories, theory of corrections, nuclear physics, economics, advertising and other real control tasks. We need a special theory and special methods for solution of these problems.

Author offers the following method, which simplifies and solves these tasks.

Key words: Optimization, impulse solutions, optimal control, aerospace..

Introduction

Optimization methos are widely used in solving of technical problems. However, there are important classes of problems where they have great difficulties in the application. For example, in problems of space travel. The fact that the operational time of conventional rocket liquid propulsion is small (minutes), while the passive time of the interplanetary flight is large (months). In the result, we can consider the rocket work as an impulse, the speed as a jump which must expend minimum fuel. In mathematics, this means: the control is at an infinity value, the phase variables have a gap, and the variables are not continuous, not differentiable. The variable culculation and the Pontryagin principle are not applicable.

In 1968 the author offered the special methods [1] (see also [2 – 3]) for solution of the difference cases the impulse regime. In book [4] he applied this method to aerospace problems. Authors of work [5] developed the impulse theory for a particular case (linear version of control) using the theory of δ-functions. But his solutions are very complex and not acceptable in many practical problems.

In the given article the author offers a simpler method for solution of these problems: he shows the known impulse problems can be reduced to the special Pontryagin problem. Their solution can be simpler than existing methods.

Statement of the problem

1. **Statement of the conventional Optimization Problem.** Assume the state of system is described by conventional differential equations:

$$I = F[x(t_1), x(t_2)] + \int_{t_1}^{t_2} f_0(t,x,u)dt, \quad \dot{x}_i = f_i(t,x,u) \quad i = 1,2,...,n, \tag{1}$$

where I is the objective function, x is n – dimentinal continuous piece-difference function of phase coordinates; u is r – dimentional piece-continuous, piece-diffrence functions of control, $a_i \leq u_i \leq b_i$, $i = 1,2,...,r$, a,b = const; t is time. End values of $x(t_1)$, $x(t_2)$ are given or mobile. F is function of the end values $x(t)$.

We must find the control u, which gives the minumum the objective function I.

In our case (impulse problem) the control (or some its components) is at infinity (a very short time), the some (or all) phase variables have the gaps, and the variables are not continuous, not differentiable. The variable calculation and Pontryagin principle are not applicable.

2. Impulse Optimization Problem. Method of Solution.

The author offer the following method for solution of impulse problems.

We enter the special constants (unknown limited values) of impulses

$$v_i \quad i=1,2,\ldots,m. \tag{2}$$

These values may be binded the contions

$$x_i^+ = x_i^- + v_i, \quad \varphi_i(t,x,u,v) = 0 \quad i=1,2,\ldots,s, \quad s<m \tag{3}$$

and limitations

$$c_{i,1} \leq v_i \leq c_{im2} \quad i=1,2,\ldots,m, \tag{4}$$

where x_i^-, x_i^+ are x_i is phase coordinate on left and on right from point of impulse (gap), $c_{i,1}, c_{i,2}$ are consts. In particule, v can be unknown constant or zero.

The optimal problem is written in fo

$$I = F[x(t_1), x(t_2)] + \int_{t_1}^{t_2} f_0(t,x,u,v)dt, \quad \dot{x}_i = f_i(t,x,u,v) \quad i=1,2,\ldots,n,$$

$$\varphi_i(t,x,u,v) = 0, \quad i=1,2,\ldots,m \tag{5}$$

Where v are unknown limited impulses (gaps). End values of $x(t_1)$, $x(t_2)$ are given or mobile.

According [2], [3], we can write the generalized functionality introdused in form

$$J = I + \alpha, \tag{6}$$

where J - the generalized functionality introduced in [2],[3] p. 42, α is so named α – function introduced in [2],[3] (function equals zero on acceptable set, for example, on curves satisfying the equations (1) – (4)).

In our case we take

$$\alpha = \int_{t_1}^{t_2} \left[\sum_{i=1}^{i=n} \lambda_i(t,x)[\dot{x}_i - f_i(t,x,u,v)] + \sum_{i=n+i}^{i=n+m} \lambda_i(t,x)\varphi_i(t,x,u,v) \right] dt, \tag{7}$$

where $\lambda(t,x)$ is an unknown vector function.

We can re-write (6) as (see [3] p.42)

$$J = I + \alpha = A + \int_{t_1}^{t_2} B \, dt, \tag{8}$$

where (for brevity repeated indices are summed):

$$A = F + \lambda_i x_i \big|_{t_1}^{t_2}, \quad B = f_0 - \left(x_j \frac{\partial \lambda_j}{\partial x_i} + \lambda_i\right) f_i - x_i \frac{\partial \lambda_i}{\partial t}, \tag{9}$$

From Theorem 3.8 [3] we get: if we find at least one solution of particular equation about λ

$$J = \inf A + \inf B, \quad \inf_{u,v}\left[f_0 - \left(x_j \frac{\partial \lambda_j}{\partial x_i} + \lambda_i \right) f_i - x_i \frac{\partial \lambda_i}{\partial t} \right], \quad \frac{\partial B}{\partial x} = 0 \tag{10}$$

for the end condition inf A, we get optimal solution.

Note, the B (9) is different from the well-known Gamiltonian. If we will take the different function λ(t,x), we will get the different conjugated system of equations ∂B/∂x = 0.

In particular, if we will get λ(t) ONLY as function t, we get the conventional Pontryagin principle of maximum

$$J = A + \int_{t_1}^{t_2} B dt, \tag{11}$$

where

$$A = F + \sum_{i=1}^{i=n}[\lambda_i(t_2)x_i(t_2) - \lambda_i(t_1)x_i(t_1)], \tag{12}$$

$$B = f_0(t,x,u,v) - \sum_{i=1}^{i=n}\lambda_i(t)f_i(t,x,u,v) - \sum_{i=n+1}^{i=n+m}\lambda_i(t)\varphi_i(t,x,u,v) \tag{13}$$

and equations

$$\dot{x}_i = f_i(t,x,u,v), \quad i = 1,2,\ldots,n \tag{14}$$

$$\dot{\lambda}_i = \frac{\partial B}{\partial x_i} \quad i=1,2,\ldots,n, \quad \bar{u} = \arg\inf_u B \quad \text{or} \quad \frac{\partial B}{\partial u_i} = 0 \quad i=,2,\ldots,r,$$

$$\bar{v} = \arg\inf_v B \quad \text{or} \quad \frac{\partial B}{\partial v_i} = 0 \quad i = r+1,2,\ldots,m, \tag{15}$$

The equations

$$\frac{\partial B}{\partial u} = 0, \quad \frac{\partial B}{\partial v} = 0 \tag{16}$$

are used only in the open area. λ_i are unknown multipliers.

Equations (11) - (16) gives the optimal trajectoris (minimum of I) of the system (5). We also must solve the boundary value problem – find such $\lambda_i(t_1)$ that to get the given $x_i(t_2)$.

The gap time t_θ and gap v inside interval ($t_1 < t_\theta < t_2$) we can also find the next way. Write the objective function in form

$$I = F[x(t_1), x(t_2)] + \Phi(t_\theta, x_\theta) + \int_{t_1}^{t_2} f_0(t,x,u)dt,$$
$$\dot{x}_i = f_i(t,x,u) \quad i = 1,2,\ldots,n, \tag{17}$$

where Φ is additional condition in t_θ (if they are given).

Write the general function as the sum of two functions in (t_1, t_θ) and $(t_1<t_\theta<t_2)$

$$J = F + \psi_2 - \psi_1 + \Phi + \psi_\theta^+ - \psi_\theta^- + \int_{t_1}^{t_\theta} Bdt + \int_{t_\theta}^{t_2} Bdt,$$

where $\psi_\theta(t_\theta) = \lambda_i x_i$, $\psi_2(t_2) = \lambda_i x_i$ $\psi_1(t_1) = \lambda_i x_i.$ (18)

In t_θ the minimal condition are

$$\inf_{t_\theta, x_\theta}[\Phi(t_\theta, x_\theta) + \psi_\theta^+(t_\theta, x_\theta) - \psi_\theta^-(t_\theta, x_\theta)],$$

$$\bar{x} = \arg\inf_x B, \quad \bar{u} = \arg\inf_u B.$$ (19)

Here up "-" and "+" are values from left and right from point t_θ.

Notes:

1. We can find in form (3) ONLY the phase coordinates which we can aproximate as the impulse (in short time we can change a large value – for example, the speed in long flight, agle of trajectory, laser excitation of atom and so on). We cannot pukes space, distance, time.

2. The λ_i of corresponding coorditate has a gap/jump in moment of impulse. The moment (time) of gap or new λ_i (at right side) we can find (in open area) from the second equation (16). We must also to check up the ends of the intersal $[t_1, t_2]$.

3. In some cases, the optimal value of gap we can find by the selection of v.

4. The λ_i of f_i are functions of t, the λ_i of φ_i are constants.

Example

Let us to consider the typical problem of space travel - transfer from one space orbit to other. Assume the space ship has circular Earth orbit having the radius r_1 and speed V_0. We want to reach the ecliptic orbit having the maximal radius $r_2 > r_1$ and spend the minimum of fuel. The liquid rocket engine works some seconds, the space flight is some months. That way we can consider the rocket flight as pulse mode which instant change speed (gap the speed). Our task is to find minimal gap of speed (minimal inpulse) $v = \Delta V$, because the minimal gap of speed is equivalent of the minimal expenditure of the rocket fuel.

Our objective function

$$I = \int_0^t \Delta V dt$$ (20)

The variables (speed V and radius r) of free space flight in the Earth gravitation field is binded by the Law of energy conservation (kinetic + potencial energy equils constant c):

$$\frac{mV^2}{2} - m\mu\left(\frac{1}{r_0} - \frac{1}{r}\right) = c, \quad \text{or} \quad V^2 = \mu\left(\frac{2}{r} - \frac{2}{r_1 + r_2}\right), \tag{21}$$

Where m is mass space ship (satellite) mass, kg; r_0 is initial radius, m; μ is gravity constant. For Earth $\mu = 3.9802 \cdot 10^{14}$ m³/s², for Sun $\mu = 1,3276 \cdot 10^{20}$ m³/s². That is elliptic orbite, r_1 is the radius of perigee; r_2 is the radius of apogee. We want to arrive from the circular orbit having V_0, the radius $r_0 = r_1$ (the point of perigee) to point of apology r_2.

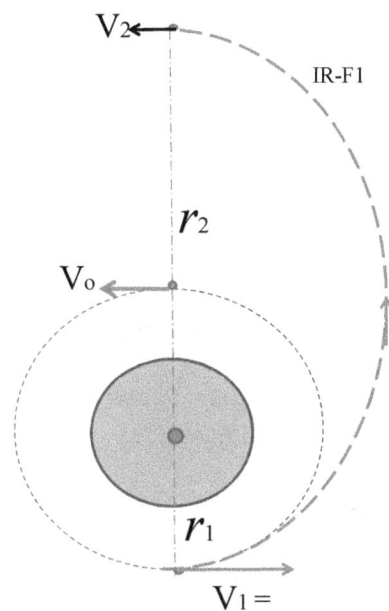

Fig.1. Orbite transver.

For elliptic orbits, the equation (21) may be re-writen in form:

$$(V_0 + \Delta V)^2 - 2\mu\left(\frac{1}{r_1} - \frac{1}{r_1 + r_2}\right) = 0 \quad \text{or} \quad V_0 + \Delta V - \sqrt{2\mu\left(\frac{1}{r_1} - \frac{1}{r_1 + r_2}\right)} = 0, \tag{22}$$

where; V_0 is speed on circular orbite having the radius r_1. The speed of circular orbit is

$$V_0 = \sqrt{\frac{\mu}{r_0}}, \quad V_1 r_1 = V_2 r_2. \tag{23}$$

Here V_2 is speed in r_2. Last equation in (23) is Law of momentum conservation free flight in the central gravitation field.

Let us the write the function B (13) for left end in right side of point t_1.

$$B = \Delta V + \lambda\left[V_1 + \Delta V - \sqrt{2\mu\left(\frac{1}{r_1} - \frac{1}{r_1 + r_2}\right)}\right], \tag{24}$$

From equation (16) we have

$$\frac{\partial B}{\partial (\Delta V)} = 1 + \lambda = 0 \quad .\tag{25}$$

The equaetion (25) together with the equations (22),(23) allow to find the λ and the speed gap ΔV:

$$\lambda = -1, \quad \Delta V = \sqrt{\frac{\mu}{r_1}}\left(\sqrt{\frac{2r_2}{r_1+r_2}}-1\right) = V_0\left(\sqrt{\frac{2r_2}{r_1+r_2}}-1\right) = V_0\left(\sqrt{\frac{2\bar{r}}{\bar{r}+1}}-1\right) = V_0\left(\frac{V_1}{V_a}-1\right), \tag{26}$$

where

$$\bar{r} = \frac{r_2}{r_1}, \quad V_a = \sqrt{\frac{V_1^2 + V_2^2}{2}}, \quad V_1 r_1 = V_2 r_2 \,. \tag{27}$$

Here V_2 is speed in apogee, V_a is average speed.

We reached the request r_2 by the first impulse. That way we don't need the additional impulse and reseach.

The formula (26) for computation ΔV is known as transfer in Gohman ellipse [6]. New is proof of optimization. The reader can solve same way the more complex inpulse (gap) problems [4].

Referances

1. Bolonkin A.A., Impulse solutions in problem of the optimal control. Prosiding of Siberian Branch of the Academy of Sciences of the USSR, series of Technical sciences. #13, Issue 3, 1968 (in Russian).

2. Bolonkin A.A., New methods of Optimizations and their application. The post-doctoral Ph.D thesis, 1969, Chapter 5, pp.120-141. Leningrad Politechnic University (in Russian).

    http://viXra.org/abs/1509.0267 Part 1, http://vixra.org/abs/1509.0265 Part2, https://www.academia.edu/s/2a5a6f9321?source=link

3. Bolonkin A.A., New methods of Optimization and their applications. Moscow High Technical University, named Bauman, 1972. 220 ps (In Russian: Новые методы оптимизации и их применение. МВТУ им. Баумана, 1972г., 220 стр., См. РГБ, Российская Государственная Библиотека, Ф-861-83/1809-6). http://vixra.org/abs/1504.0011 v4. , https://www.academia.edu/11054777/ v.4

4. Bolonkin AA., "Non Rocket Space Launch and Flight". Elsevier, 2005. 488 pgs. ISBN-13: 978-0-08044-731-5, ISBN-10: 0-080-44731-7 .

    https://archive.org/details/Non-rocketSpaceLaunchAndFlightv.3 , (v.3), http://vixra.org/abs/1407.0174 ,

5. Дыхта В.А., Самсонюк О.Н., Оптимальные импульные управления с приложениями. М. Наука, ФМ, 2000, (in Russian).

6. Walter Hohmann. Die Erreichbarkeit der Himmelskörper. — Verlag Oldenbourg in München, 1925. — ISBN 3-486-23106-5.

30 September 2015.

General references of some Bolonkin publications

List #1 of some Bolonkin's publication in 1958-1972 (in Russian).

1. Болонкин А.А., Проектирование самолета с подводным крылом. Дипломный проект. Самолетостроительное ОКБ О.К. Антонова. Киев, п/я 4. 1958г.
2. Болонкин А.А., Исследование движения самолета АН-12 на разбеге. Отчет ОКБ п/я 4, 1959г.
3. Болонкин А.А., Исслеования по выбору оптимальных параметров самолета с вертикальным взлетом и посакой. Отчет ОКБ п/я 4, 1959г.
4. Болонкин А.А., Исследования по выбору оптимальных параметров высотного самолета. Отчет ОКБ п/я 4, 1959г.
5. Болонкин А.А., Расчет летных данных самолетов высокой тяговооруженности. Отчет ОКБ п/я 4, 1959г
6. Болонкин А.А., Метод оценки эффективности управления пограничным слоем на самолете. Труды КВИАВУ, вып.72, 1960г.
7. Болонкин А.А., Теория полета летающих моделей. Москва, Издательство ДОСААФ, 1962г, 327 стр. http://www.twirpx.com/file/138441/
8. Болонкин А.А., Оптимизация траекторий. Отчет МАИ , каф.109. 1962г.
9. Болонкин А.А., Некоторые вопросы исслеования траекторий летательных аппаратов. МАИ. Отчет по теме 0411, 1963г.
10. Болонкин А.А., Приближенный метод решения задач оптимального управления. Отчет МАИ по теме 0411, 1964г.
11. Болонкин А.А., Оптимизация параметров в вариационных задачах.ДАН УССР, №5. 1964г.
12. Болонкин А.А., Принцип расширения и условие Якоби вариационного исчисления. ДАН УССР, №7.1964г
13. Болонкин А.А., Вариационное исчисление и функциональное уравнение Беллмана. ДАН УССР, №10, 1964г.
14. Болонкин А.А., Метод решения оптимальных задач. Сб. «Сложные системы управления». Киев, Наукова Думка, 1965г., стр. 34-67.
15. Болонкин А.А., Достаточные условия в разрывных вариационных задачах с ограниченным управлением. Приложение к статье «Оптимизация траекторий многоступенчатых летательных аппаратов». В сб. «Исследования по динамике полета», Машиностроение. 1965г., стр. 44-78.
16. Болонкин А.А., Оптимизация траекторий многоступенчатых летательных аппаратов. Сб. «Исследования по динамике полета»., Москва, Машиностроение, 1965г., стр. 20-78.
17. Болонкин А.А., Особые, скользящие и импульсные режимы в задачах динамики полета. Сб. «Сложные системы управления», Киев, Наукова Думка, 1965г., стр.68-90.
18. Болонкин А.А., Исследование динамики старта самолета с вертикальным взлетом. Сб. «Исследования по динамике полета»., Москва, Машиностроение, 1965г., стр. 119-147.
19. Болонкин А.А., Теория полета аппаратов с управляемой радиальной силой. Сб. «Исследования по динамике полета»., Москва, Машиностроение, 1965г., стр. 79-118.
20. Болонкин А.А., Оптимизация траекторий многоступенчатых ракет. Кандидатская диссертация, МАИ, 1965г.
21. Закрытие отчеты по исследованию ракетных двигателей. Ракетное ОКБ Глушко. Химки, 1966г
22. Болонкин А.А., О разрешимости краевых задач оптимального управления.Труды академии им. Н.Е.Жуковского, вып. 1131, 1966, стр.103-128.
23. Болонкин А.А., О решении общей задачи линейного оптимального быстродействия с одним управлением. Прикладная механика, т.4. вып.4, 1968г., стр. 111-121 (получены редакцией 18.4.66г).

24. Болонкин А.А., Специальные экстремали в задачах оптимального управления. Тезисы докладов и сообщений на Всесоюзном межвузовском симпозиуме по прикладной математике и кибернетике, Горький, 1967г.
25. Болонкин А.А., Специальные экстремали. МАТИ, Отчет, 1967г
26. Болонкин А.А., Особые решения в задачах аналитического конструирования оптимальных регуляторов.МАТИ, Отчет. 1967г.
27. Болонкин А.А., Геворкян А.М., Полипаев, Мето интегрированной обработки окументации, МАТИ, Отчет по теме ,1968г.
28. Болонкин А.А., Импульсные решения в задачах оптимального управления. Известия Сибирского Отеления Акаемии Наук СССР. Серия технических наук, №13, вып.3, 1968г.
29. Болонкин А.А., Специальные экстремалив задачах оптимального управления. Журнал «Техническая кибернетика», «2, 1969г.
30. Болонкин А.А., Решение дискретных заач оптимального управления на основе общего принципа минимума. Сб. Вычислительная и прикланая математика. КГУ, вып.7, 1969г.
31. Болонкин А.А., Об одном методе решения оптимальных заач. Известия СО АН СССР, вып. 2, №8, июнь 1970г.
32. Болонкин А.А., Об одном подходе к решению оптимальных задач. Сб. Вычислительная и прикладная математика, КГУ, вып.12, 1970г.
33. Болонкин А.А., О решении оптимальных заач. Сб. «Математические вопросы управления произвоством», вып. 3, м., 1971г.
34. Болонкин А.А., Методы решения краевых задач теории оптимального управления. Журнал «Прикладная механика», т.4, вып.6, 1971г.
35. Болонкин А.А., Новые методы оптимизации и их применение. Диссертация на соискание ученой степени доктора технических наук. Москва. 1969г.285 стр. См. Автореферат диссертации на соискание ученой степени доктора технических наук. Москва, ЛПИ, 1971г., 28 стр. http://viXra.org/abs/1503.0081.
36. Болонкин А.А., Новые методы оптимизации и их применение. Краткий конспект лекций по курсу Теория оптимальных систем. МВТУ им. Баумана, 1972г., 220 стр. (См. РГБ, Российская Государственная Библиотека, Ф-861-83/1869-6). http://viXra.org/abs/1502.0137, v/3, https://archive.org/details/BookOptimization3InRussianInWord20032415_201502 v.3.

List #1 of some Bolonkin's publications in 1958-1972 (Titles are translated in English).

[1] Bolonkin, A.A., (1958a). Design of Hydro-Aircraft with Underwater Wing. *Report of Aircraft State Construction Bureau named Antonov*, Kiev, Ukraine, 1959 (in Russian), 120 p.

[2] Bolonkin, A.A., (1959a). Investigation of Aircraft AN-12 in Take-off. *Report of Aircraft State Construction Bureau named Antonov*, Kiev, Ukraine, 1959 (in Russian), 60 p.

[3] Bolonkin, A.A., (1959b). Research of Optimal Parameters of VTOL Aircraft, *Report of Aircraft State Construction Bureau named Antonov*, Kiev, 1959 (in Russian), Part 1 40 p., part 2 35 p.

[4] Bolonkin, A.A., (1959c). Research of Optimal Parameters of High Altitude Aircraft. *Report of Aircraft State Construction Bureau named Antonov*, Kiev, Ukraine, 1959 (in Russian), 60 p.

[5] Bolonkin, A.A., (1959d). Computation of High Thrust Aircraft with unsteady polar. *Report of Aircraft State Construction Bureau named Antonov*, Kiev, Ukraine, 1959 (in Russian), 50 p.

[6] Bolonkin, A.A., (1960a). Method of Estimation the Control of Aircraft Interface. *Presiding of Kiev High Engineering Aviation Military School* (KVIAVU), Kiev, Ukraine, Issue #72, 1960.

[7] Bolonkin, A.A., (1962a), *Theory of Flight Models*, Moscow, Association of Army, Air Force, and NAVY, 328p. 1962, (in Russian).

[8] Bolonkin, A.A., (1964a). Optimization of parameters of variation problems (Ukrainian. Russian and English summaries). *Dopoviti Akad. Nauk Ukrain. RSR,* 1964, #5, p. 580-582. Math.Rev. #6352.

[9] Bolonkin, A.A., (1964b) The calculus of variations and a functional equation of Bellman, and an interpretation of Langrange's undetermined multipliers. (Ukrainian. Russian and English summaries). *Dopovidi Akad. Nauk Ukrain., RSR* 1964, #10, p. 1290-1293. M.R. #5136.

[10] Bolonkin, A.A., (1964c). The extension principle and the Jacobi condition of the variation calculus. (Ukrainian. Russian and English summaries). *Dopovidi Akad. Nauk Ukrain. RSR* 1964. #7. p. 849-853. M.R. #5117.

[11] Bolonkin, A.A., (1965a), "Theory of Flight Vehicles with Control Radial Force". Collection *Researches of Flight Dynamics*, Mashinostroenie Publisher, Moscow, pp. 79–118, 1965, (in Russian). Intern.Aerospace Abstract A66-23338# (Eng).

[12] Bolonkin, A.A., (1965b), Investigation of the Take off Dynamics of a VTOL Aircraft. Collection *Researches of Flight Dynamics*. Moscow, Mashinostroenie Publisher, 1965, pp. 119-147 (in Russian). International Aerospace Abstract A66-23339# (English).

[13] Bolonkin, A.A., (1965c), Optimization of Trajectories of Multistage Rockets. Collection *Researches of Flight Dynamics*. Moscow, 1965, p. 20-78 (in Russian). International Aerospace Abstract A66-23337# (English).

[14] Bolonkin, A.A., (1965d). A method for the solution of optimal problems (Russian). Collection *Complex Systems Control*, pp.34-67. Naukova Dumka, Kiev, 1965. M.R. #5535.

[15] Bolonkin, A.A., (1965e). Special, Sliding, and Impulse Regimes in Problems of Flight Dynamics (Russian). Collection *Complex Systems Control*, pp.68-90. Naukova Dumka, Kiev, 1965. M.R. #5535.

[16] Bolonkin, A.A., (1966a). Boundary-value problems of Optimal Control. Military *Aviation Engeenering Academy (VVIA) named Zhukovskii,* Issue #1131, 1966, p. 103-128. (Russian).

[17] Bolonkin, A.A., (1967a). Special extreme, *Report presented to Symposium of Applied Mathematics*, Gorkii, USSR, 1967. (Russian).

[18] Bolonkin, A.A., (1968a). Impulse solution in control problems. (Russian). *Izv. Sibirsk. Otdel. Akad. Nayk USSR*, 1968, No. 13, p. 63-68. M.R. 7568. (Russian).

[19] Bolonkin, A.A., (1968b). Solution of Problem the Linear Optimal Control with one Control, Journal *"Prikladnaya Mechanica"*, Vol. 4, #4, 1968, p.111-121. (Russian).

[20] Bolonkin, A.A., (1969a). Solution of discrete problems of optimal control on the basis of a general minimum principle (Russian. English summary). *Vycisl. Prikl. Mat. (Kiev) Vyp. 7 (1969)*, 121-132. Mathematical Review 771.

[21] Bolonkin, A.A., (1969b), Special extreme in optimal control. Akademia Nauk USSR, Izvestiya. Tekhnicheskaya Kibernetika, No 2, Mar-Apr.,1969, p.187-198. See also *English translation in* Engineering Cybernetics, # 2, Mar- Apr.1969, p.170-183, (English).

[22] Bolonkin, A.A., (1970a). A certain method of solving optimal problems. *Izv. Sibirsk. Otdel. Akad. Nauk SSSR*. 1970, no.8, p. 86-92. M.R. #6163.

[23] Bolonkin, A.A., (1970b). A certain approach to the solution of optimal problems. (Russian. English summary). *Vycisl. Prikl. Mat.* (Kiev). Vyp. 12 (1970), p. 123-133. M.R. #7940.

[24] Bolonkin, A.A., (1971a), *Solution Methods for boundary-value problems of Optimal Control Theory*. Translated from Prikladnaya Mekhanika, Vol. 7, No 6, 1971, p. 639-650, (in English).

[25] Bolonkin, A.A., (1971b). Solution of Optimal Problems. Collection *"Mathematical Problems of Production Control"*, Moscow State University (MGU), Issue #3, 1971, p. 55-67.

[26] Bolonkin, A.A., (1972a), *New Methods of Optimization and their Applications*, Moscow Highest Technology University named Bauman, 1972, p.220 (in Russian).

List 5.3s of Bolonkin's publications in 2007-2016 (Loading is free):

List 5.3 contains more detail of Bolonkin's publications in 2007-2015 (Loading is free).:

The List 5.2 contains of Bolonkin's scientific researches, works, articles and books written or published in 2007-2015 and free links to them. http://viXra.org/abs/1604.0304 , 4 22 16; https://archive.org/details/List5.2OfBolonkinPublication42216 ,

1. **List 5.1 of Bolonkin's publications in 2007-2014 (Loading is free)**: https://archive.org/details/List5OfBolonkinPublications, http://intellectualarchive.com #1392. http://samlib.ru/editors/b/bolonkin_a_a/list5bolonkinspublicationin2007-2014.shtml , https://www.academia.edu/9251433/, Ref. 8386086, 4 1 15
2. AIAA. American Institute Aeronautics and Astronautics (>40 Bolonkin's scientific reports). http://arc.aiaa.org/action/doSearch?AllField=Bolonkin
3. Archive of Cornel University (USA): (45 Bolonkin's scientific works). http://arxiv.org/find/all/1/au:+Bolonkin/0/1/0/all/0/1
4. Archives (USA): Search: "Bolonkin" (>86 Bolonkin's scientific works). https://archive.org/search.php?query=Bolonkin http://www.archive.org .
5. Intellectual Archive, (>40 Bolonkin's scientific works). http://intellectualarchive.com/?link=find# . Search: Physics, Bolonkin
6. Bolonkin's WEB http://Bolonkin.narod.ru/p65.htm .
7. http://www.scribd.com, https://www.scribd.com/Bolonkin (it may be temporary blocked) (>98 Bolonkin's scientific works).
8. Archive Vixra. http://vixra.org/search?domains=vixra.org&q=Bolonkin&client=pub-9708849425281176&forid=1&ie=ISO-8859-1&oe=ISO-8859-1&cof=GALT%3A%23008800%3BGL%3A1%3BDIV%3A%23ffffff%3BVLC%3A663399%3BAH%3Acenter%3BBGC%3Affffff%3BLBGC%3Affffff%3BALC%3A008800%3BLC%3A008800%3BT%3A000000%3BGFNT%3A008800%3BGIMP%3A008800%3BFORID%3A11&hl=en&sitesearch=vixra.org%2Fpdf (45)
9. Academia.edu https://independent.academia.edu/AlexanderBolonkin/Papers (>19)
10. General Science Journal, (>12 scientific works). http://gsjournal.net/Science-Journals-Papers/Author/1481/Alexander,%20Bolonkin
11. Cheapest Bolonkin books: http://www.lulu.com/shop/search.ep?keyWords=Bolonkin&sorter=relevance-desc
12. Болонкин в Энциклопедии «Ученые России» http://www.famous-scientists.ru/14910/
13. Bolonkin in English Wikipedia http://en.wikipedia.org/wiki/Alexander_Bolonkin
14. Storige: https://disk.yandex.ru/client/narod
15. Add to scribd: https://www.scribd.com/search-documents?query=Bolonkin
16. https://www.lulu.com/shop/search.ep?keyWords=Bolonkin&type=&pn=2
17. The cheapest copies in Lulu (0 - $29): https://www.lulu.com/shop/search.ep?keyWords=Bolonkin&type=&pn=2

Books:

1) **Human Immortality and Electronic Civilization**. Electronic book, 1991. WEB: http://Bolonkin.narod.ru, http://bolonkin.narod.ru/Book_Immortality_in_English.htm http://bolonkin.narod.ru/Book_Immortality_in_Russian.htm .??? http://narod.ru/disk/13292584000/Book_Immortaity_in_Russian_6_4_07.doc.html

2) **Human Immortality and Electronic Civilization**. Lulu, USA, 3rd edition (in English), 2007. http://www.archive.org/details/HumanImmortalityAndElectronicCivilization , http://Vixra.org/abs/0709.0001 , http://Vixra.org/pdf/0709.0001v1.pdf , https://www.academia.edu/11001577/Human_Immortality_and_Electronic_Civilization_USA_New

3) **Бессмертие людей и электронная цивилизация** (in Russian). USA, Lulu, 3-е издание 2007, http://www.archive.org/details/HumanImmortalityAndElectronicCivilizationInEussian, http://viXra.org/abs/1309.0189 OK, (Russian), http://intellectualarchive.com/?link=find#detail №1140, https://www.scribd.com/search-documents?query=Bolonkin ?, http://www.scribd.com/doc/24052811 ?.

4) **Записки советского политзаключенного**. 1991 (in Russian). 70 стр., http://vixra.org/abs/1309.0187, http://intellectualarchive.com/?link=find#detail #1142, http://www.archive.org/details/MemoirsOfSovietPoliticalPrisonerinRussian, https://www.scribd.com/search-documents?query=Bolonkin , http://www.scribd.com/doc/24053537

5) **Memories of Soviet Political Prisoner.** Translation from Russian. Lulu, 1995. http://viXra.org/abs/1309.0188, http://www.archive.org/details/MemoirsOfSovietPoliticalPrisoner, http://intellectualarchive.com/?link=find#detail #1141, https://www.academia.edu/11001770/Memoirs_of_Soviet_Political_Prisoner_New_York_1991 , https://www.scribd.com/search-documents?query=Bolonkin, http://www.scribd.com/doc/24053855,

6) **"Non Rocket Space Launch and Flight".** Elsevier, 2006. 488 pgs. ISBN-13: 978-0-08044-731-5, ISBN-10: 0-080-44731-7 . https://archive.org/details/Non-rocketSpaceLaunchAndFlightv.3 , (v.3) http://vixra.org/abs/1407.0174 , http://www.archive.org/details/Non-rocketSpaceLaunchAndFlight , (corrected), http://www.scribd.com/doc/203941769/Non-Rocket-Space-Launch-and-Flight-v-3 , http://www.scribd.com/doc/24056182 (graph). http://www.scribd.com/doc/202159078/Non-Rocket-Space-Launch-and-Flight-2-nd-Edition (graph). http://www.twirpx.com/file/1296604/ , https://www.academia.edu/11055944/Non_Rocket_Space_Launch_and_Flight_Graph_V.3 http://narod.ru/disk/13288386000/Book%20Non%20Rocket%20v2.doc.html .

7) **"New Concepts, Ideas, Innovations in Aerospace, Technology and the Human Sciences"**, NOVA, 2006, 510 pgs. ISBN-13: 978-1-60021-787-6. http://viXra.org/abs/1309.0193, http://www.archive.org/details/NewConceptsIfeasAndInnovationsInAerospaceTechnologyAndHumanSciences http://www.scribd.com/doc/24057071 ?, https://independent.academia.edu/AlexanderBolonkin/Papers ? http://narod.ru/disk/13289623000/Bolonkin%20Monograph-P%20corrected%2011%2020%202007.doc.html ,16Mb.

7) **"Macro-Projects: Environments and Technologies"**, NOVA, 2007, 536 pgs. ISBN 978-1-60456-998-8. http://www.archive.org/details/Macro-projectsEnvironmentsAndTechnologies, http://viXra.org/abs/1309.0192 , http://www.scribd.com/doc/24057930 , https://independent.academia.edu/AlexanderBolonkin/Papers,Storige: http://narod.ru/disk/13292420000/Book_Macro_Projects_for_Internet%209%2018%2009.doc.html

8) **"New Technologies and Revolutionary Projects"**, Scribd, 2008, 324 pgs, http://www.archive.org/details/NewTechnologiesAndRevolutionaryProjects, https://www.scribd.com/search-documents?query=Bolonkin , http://www.scribd.com/doc/32744477 ,

9) **«Жизнь. Наука. Будущее»** (биографические очерки, исследования и инновации), Россия, ПГУ, 2010, 286 pgs, 23 Мб., ISBN: 978-1-300-49164-4, http://viXra.org/abs/1309.0204, http://www.twirpx.com/file/1592630/ or http://www.archive.org/details/Life.science.futureinRussian..., https://www.scribd.com/search-documents?query=Bolonkin , http://www.scribd.com/doc/45901785

10) **LIFE. SCIENCE. FUTURE (Biography notes, researches and innovations).** Lambert, 2010, 208 pgs. 16 Mb.ISBN: 978-3-8473-0839-3 . http://vixra.org/pdf/1309.0205v1.pdf http://viXra.org/abs/1309.0205, http://www.lulu.com, search "Bolonkin"; http://www.archive.org/details/Life.Science.Future.biographyNotesResearchesAndInnovations,

https://www.scribd.com/search-documents?query=Bolonkin, http://www.scribd.com/doc/48229884, or http://www.publishamerica.net/sc/productsearch.cgi?search_field=Bolonkin&storeid=*1ed736148e14b9de8c3184b7f08fb4a634 , or http://www.amazon.com/s/ref=nb_sb_noss?url=search-alias%3Dstripbooks&field-keywords=Bolonkin&x=12&y=19 .

11) **Universe, Human Immortality and Future Human Evaluation.** Elsevier. 2010r., 124 pages, 4.8 Mb. ISBN-10: 0124158013, ISBN-13: 978-0124158016
http://www.archive.org/details/UniverseHumanImmortalityAndFutureHumanEvaluation, ??
http://viXra.org/abs/1207.0020, ?? http://intellectualarchive.com/?link=find#detail, ??
https://www.scribd.com/search-documents?query=Bolonkin
http://www.scribd.com/doc/52969933/ bloked 7 26 12.
https://independent.academia.edu/AlexanderBolonkin/Papers .??

12) **Human immortality and Electronic Civilization,** Publish America, Baltimore, USA, 2010,140 ps.
ISBN: 978-1-4489-3969-5, 140 pages, 5.5 x 8.5, $9.95.
http://www.publishamerica.net/sc/productsearch.cgi?search_field=Bolonkin&storeid=*1ed736148e14b9de8c3184b7f08fb4a634 or http://www.amazon.com/s/ref=nb_sb_noss?url=search-alias%3Dstripbooks&field-keywords=Bolonkin&x=12&y=19

13) **Memoirs of Soviet Political Prisoner,** Publish America, Baltimore, USA, 2010,108 ps.
ISBN: 978-1-4489-4414-9, 108 Pages, 5.5 x 8.5, $9.95.
http://www.publishamerica.net/sc/productsearch.cgi?search_field=Bolonkin&storeid=*1ed736148e14b9de8c3184b7f08fb4a634

14) **LIFE. SCIENCE. FUTURE (Biography notes, researches and innovations).** Publish America, Baltimore, USA,2010,208 pgs.16 Mb. ISBN: 978-1-4512-7983-2, 306 Pages, 6x9, $15.95.
http://www.archive.org/details/Life.Science.Future.biographyNotesResearchesAndInnovations,
http://www.lulu.com search "Bolonkin" or
http://www.lulu.com/shop/search.ep?keyWords=Bolonkin&sorter=relevance-desc .
http://www.publishamerica.net/sc/productsearch.cgi?search_field=Bolonkin&storeid=*1ed736148e14b9de8c3184b7f08fb4a634 or http://www.amazon.com/s/ref=nb_sb_noss?url=search-alias%3Dstripbooks&field-keywords=Bolonkin&x=12&y=19 . or http://www.scribd.com/doc/48229884,

15) **Femtotechnologies and Revolutionary Projects.** Lambert, USA, 2011. 538 p. 16 Mb.
ISBN:978-3-8473-0839-0. http://viXra.org/abs/1309.0191,
http://www.archive.org/details/FemtotechnologiesAndRevolutionaryProjects ,
https://independent.academia.edu/AlexanderBolonkin/Papers

16) **Life and Science.** Lambert Academic Publishing, Germany, 2011, 205 pgs. ISBN: 978-3-8473-0839-3.
http://www.archive.org/details/Life.Science.Future.biographyNotesResearchesAndInnovations,
https://www.academia.edu/11001078/LIFE._SCIENCE._FUTURE_Biography_notes_researches_and_innovatis_
https://www.scribd.com/search-documents?query=Bolonkin , http://www.scribd.com/doc/48229884,

17) **Universe, Human Immortality and Future Human Evaluation,** Elsevier, 2011.
http://www.archive.org/details/UniverseHumanImmortalityAndFutureHumanEvaluation, ?
http://intellectualarchive.com/?link=find#detail, http://viXra.org/abs/1207.0020 ?
https://independent.academia.edu/AlexanderBolonkin/Papers

18) **Innovations and New Technologies (v.1).** Scribd, 30/7/2013. 309 pgs. 8 Mb.
http://viXra.org/abs/1307.0169,
http://archive.org/details/InnovationsAndNewTechnologies,http://intellectualarchive.com/ , #1115,
http://www.scribd.com/doc/157098739/Innovations-and-New-Technologies-7-9-13 ,
https://www.academia.edu/11000128/Innovations_and_New_Technologies_v.2_ (new)

19) **Femtotechnologies and Innovative Projects**, NY., Lulu (www.lulu.com, ID 12837818), 2011, 518 ps., ISBN 978-1-105-64111-4, ISBN 978-1-300-48671-8. http://viXra.org/abs/1309.0191 (OK).
http://archive.org/details/FemtotechnologiesAndRevolutionaryProjects_873,
http://www.twirpx.com/file/1593422/.

20) **Femtotechnologies and Revolutionary Projects**, Lambert (Academic Publishing), UK., 2010, 530ps.,ISBN: 978-3-8473-2229-0.
http://archive.org/details/FemtotechnologiesAndRevolutionaryProjects_873 ,
http://www.twirpx.com/file/1593422/ .
https://labs.inspirehep.net/submit/literature/11822

21) Slides for book "Electronic Immortality",

http://www.scribd.com/doc/162847655/Immortality-1-2-3-4-5-6-7-8-9-10-11

22) **Innovations and New Technologies (v2).** Lulu, 2013. 465 pgs. 10.5 Mb, ISBN: 978-1-312-62280-7.
https://archive.org/details/Book5InnovationsAndNewTechnologiesv2102014/,
http://intellectualarchive.com/ Search: Bolonkin, http://www.twirpx.com/file/1593437/ ,
http://samlib.ru/editors/b/bolonkin_a_a/innovationsandnewtechnologies.shtml
https://www.academia.edu/11058394/Innovations_and_New_Technologies_v.2_USA_Lulu_2013

23) Болонкин А. А. **Теория полета летающих моделей**. — М.: Изд-во ДОСААФ, 1962. 327 с.
http://www.twirpx.com/file/138441/ Скачано в Download DJVU (Temp),

24) Болонкин А.А., **Новые методы оптимизации и их применение**. МВТУ им. Баумана, 1972г., 220 стр. (См. РГБ, Российская Государственная Библиотека, Ф-861-83/1809-8).
http://vixra.org/abs/1504.0011 **v4.** http://viXra.org/abs/1502.0137 **v3;**
http://viXra.org/abs/1502.0055 v2; http://viXra.org/abs/1501.0228, (v1, old) ,
https://archive.org/details/BookOptimization3InRussianInWord20032415 v2,
https://archive.org/details/BookOptimization3InRussianInWord20032415_201502 v3,
https://archive.org/details/BookOptimizationInRussian (old),
http://intellectualarchive.com/ (v2) again 2 6 15 **?, v3 2 16 15, (v.4) 3 17 15 (must be<7Mb)**
http://www.twirpx.com/file/1592607/ 2 6 15 загрузил v2.
http://www.twirpx.com/file/1605604/?mode=submit v3, загрузил 2 16 15
https://www.academia.edu/11054777/ v.4

25) Болонкин А.А., **Обыкновенный коммунизм**. Lulu, USA, 2014. 2411Кв. ISBN 978-1-312-95386-4.
https://archive.org/details/BookOrdinaryCommunism2122814FromPenskyAfter2NdCorr
http://intellectualarchive.com/ #1437 , http://www.twirpx.com/file/1592636/
http://samlib.ru/editors/b/bolonkin_a_a/bookordinarycommunism102414.shtml

26) Болонкин А.А., **Новые методы оптимизации и их примеение в задачах динамики управляемых систем.** Автореферат диссертации на соискание ученой степени доктора технических наук. Москва, ЛПИ, 1971г., 28 стр.
http://viXra.org/abs/1503.0081, 3 11 15.
http://www.twirpx.com , https://archive.org ?(не загружается?>7 Mb?)
http://samlib.ru/editors/b/bolonkin_a_a/ , http://intellectualarchive.com/
https://independent.academia.edu/AlexanderBolonkin/Papers, #1488.

27) **List #1 Bolonkin's publications in 1965-1972.**(in Russian).
https://archive.org/details/No1119651972

28) **The Development of Soviet Rocket Engines (For Strategic Missiles).** 1991, Delphic Associates, Inc., USA, 133 ps., ISBN 1-55831-130-0. (See Amazon)

29) Chapter: **Aviation, Motor and Space Designs**. In Collection: "Emerging Technology in the Soviet Union". Part 1: 1990, Delphic Associates, Inc., USA, pp.33-80., ISBN 1-55831-117-1.

30) Chapter: **Aviation Designs and Other Projects**. In Collection: "Emerging Technology in the Soviet Union". Part 2: 1990, Delphic Associates, Inc., USA, 43 p.., ISBN 1-55831-117-1.

31) **Optimal Trajectories of Air and Space Vehicles.** For BP.
https://archive.org/details/ArticleOptimalTrajectoriesOfAirAndSpaceVehicles4115,
https://independent.academia.edu/AlexanderBolonkin/Papers,
http://intellectualarchive.com/?link=find#detail #1498, http://vixra.org/pdf/1504.0015v1.pdf

32) **Bolonkin in Russian WIKI** in Word-93
https://archive.org/details/BolonkinInWIKIInCodesWord932215 ,
http://intellectualarchive.com/?link=find#detail #1524.

33) **Докторская диссертация** А.Болонкина: Новые метоы оптимизации и их применение в задачах динамики управляемых систем. ЛПИ 1969г.
https://drive.google.com/file/d/0BzlCj79-4Dz9YTJOUHVhR1FZUVE/view?usp=drive_web Dissertation Optimization 1-2 9 30 15.doc,
http://viXra.org/abs/1511.0214 ,
   http://viXra.org/abs/1509.0267 Part 1, http://vixra.org/abs/1509.0265 Part2
   https://archive.org/details/NewMethodsOfOptimizationAndItsApplication.Part1inRussian
   https://www.academia.edu/s/2a5a6f9321?source=link , http://www.twirpx.com,

34) **Краткая информация о Болонкине**. http://viXra.org/abs/1512.0229,
   https://archive.org/details/122814_201512 , В кодах ВИКИ https://archive.org/details/Index_of_/7/items/BolonkinInWIKIInCodesWord932215/ . (?),
   https://www.academia.edu/s/acc8c35402?source=link ,
   https://www.scribd.com/doc/292298227/%D0%A1%D1%82%D0%B0%D1%82%D1%8C%D1%8F-%D0%BE-%D0%B4-%D1%82-%D0%BD-%D0%90%D0%BB%D0%B5%D0%BA%D1%81%D0%B0%D0%BD%D0%B4%D1%80%D0%B5-%D0%91%D0%BE%D0%BB%D0%BE%D0%BD%D0%BA%D0%B8%D0%BD%D0%B5-2-1-27-13-For-Merge
   http://intellectualarchive.com #1647. http://newconcepts.club/website/articles/1500.html ,

35) Болонкин А. А., **Об одном методе решения оптимальных задач**. Известия СО Академии наук СССР, вып.2, № 8, июнь 1970 г. http://www.twirpx.com/file/1837179/ ,
   http://viXra.org/abs/1512.0357 , http://vixra.org/pdf/1512.0357v1.pdf ,
   https://www.academia.edu , https://archive.org/download/ArticleMethodSolutionOfOptimalProblemsByBolonkin

36) **Приговор Мосгорсуда правозащитнику Александру Болонкину**. 23 November 1973.
   https://archive.org/details/abolonkin_gmail , http://intellectualarchive.com #1657,
   https://ia601509.us.archive.org/17/items/abolonkin_gmail/%d0%9f%d1%80%d0%b8%d0%b3%d0%be%d0%b2%d0%be%d1%80%20%d0%a5%d0%be%d1%80%d0%be%d1%88%d0%b0%d1%8f%20%d0%ba%d0%be%d0%bf%d0%b8%d1%8f.pdf , http://www.twirpx.com/file/1836584/ ,
   www.IntellectualArchive.com/getfile.php?file=H0HaP3n8IKT&orig_file =Приговор Хорошая копия Word 2003.doc

37) **Small Non-Expensive Electric Cumulative Thermonuclear Reactors.** Lulu, March 2017. ISBN 978-1-312-62280-7. http://www.twirpx.com/file/2188829/?note=added-unapproved ,
   https://archive.org/details/Book2ElectricImpulseThermoReactorsAfterOlga21216165
   www.IntellectualArchive.com #1829, 3 20 17. http://viXra.org/abs/1703.0199,
   https://narod.academia.edu/AlexanderBolonkin ,

38) **Preon Interaction Theory and Model of Universe**. USA, Lulu, March 2017.
   ISBN 978-1-365-79994-6. Lulu 3 5 17 , http://viXra.org/abs/1703.0200 , 3 20 17.
   .https://archive.org/download/BookPreonTheoryAndUniverseFromLulu3517 ,
   https://narod.academia.edu/AlexanderBolonkin , www.IntellectualArchive.com #1828,
   http://www.twirpx.com/file/2188818/?note=added-unapproved ,
   www.IntellectualArchive.com/getfile.php?file=ZjORQQa9jFN&orig_file=Book_Preon_Theory_and_Universe_from

39) **Wind Energy. Electron Jet Generators and Propulsion**. USA, Lulu, 2017, 142p.,
   ISBN 978-1-365-84732-5. http://viXra.org/abs/1704.0212 4/16/17, <info@archive.org>;
   https://archive.org/details/BookWindEnergy32117.docx (3.68 MB) ???
   www.IntellectualArchive.com/#1839 ,

https://narod.academia.edu/Papers/Upload?cb=https%3A%2F%2Fnarod.academia.edu%2F&section=Papers, OK,
http://www.IntellectualArchive.com/getfile.php?file=cLefGv57VQI&orig_file=Book Wind Energy 3 21 17.docx

40) **Universal Optimization and its Application** (English excerpts).
http://viXra.org/abs/1704.0193. 4 14 17.
https://archive.org/details/BookOptimizarionChapter1241317 . English

41) **Popular Review of new Concepts, Ideas and Innovations in Space Launch and Flight.** USA, Lulu, 2017, 160 pg. Vixra 5 22 17. http://viXra.org/abs/1705.0333 , https://archive.org/details/PopularBookReviewNewIdeasInSpace

42) **Preon Interaction Theory and Model of Universe (v2).** Lulu, 8/25/2017. 105 ps. ISBN 978-1-387-18552-8.

43) **High-rise and Space Towers** (Mast, Space Elevator, Motionless Satellities). Lulu, 2017, 179 ps. ISBN: 978-1-387-18533-7

Articles (2009-2016) (loading is free):

1) **Femtotechnology: Nuclear AB-Material with Fantastic Properties.**
American Journal of Engineering and Applied Science, Vol. 2, #2, 2009, pp.501-514. Presented as paper AIAA-2009- 4620 to 7th Annual International Energy Convention Conference, 2-5 August 2009, Denver, CO, USA. http://viXra.org/abs/1309.0201, http://www.scribd.com/doc/24046679/ . Book "Femtotechnology", USA, Lulu, 2009.

2) **Femtotechnology: Design of the Strongest AB-Matter for Aerospace.** Presented as paper AIAA-2009-4620 to 45 Joint Propulsion Conference, 2-5 August, 2009, Denver CO, USA. See also closed paper AIAA-2010-1556 in 48 Aerospace Meeting, New Horizons, 4 – 7 Jan, 2010, Orlando, FL, USA. Global Scien Journal http://gsjournal.net/Science-Journals/Research%20Papers-Engineering%20%28Applied%29/Download/5713 , Journal of Aerospace Engineering, Oct. 2010, Vol. 23, No. 4, pp.281-292. http://viXra.org/abs/1401.0173 , http://www.archive.org/details/FemtotechnologyDesignOfTheStrongestAb-matterForAerospace, http://intellectualarchive.com #1362 . http://gsjournal.net/Science-Journals/Essays/View/5713 , https://www.academia.edu/14515249/FEMTOTECHNOLOGY_THE_STRONGEST_AB-MATTER_FOR_AEROSPACE

3) **Converting of Any Matter to Nuclear Energy by AB-Generator**
American Journal of Engineering and Applied Science, Vol. 2, #4, 2009, pp.683-693. Presented as paper AIAA-2009-5342 in 45 Joint Propulsion Conferencees, 2-5 Augest, 2009, Denver, CO. http://viXra.org/abs/1309.0200, https://www.scribd.com/search-documents?query=Bolonkin, https://www.academia.edu/14515398/Converting_of_Matter_to_Nuclear_Energy

4) **Converting of any Matter to Nuclear Energy by AB-Generator and Aerospace**
Journal of Energy Storage and Conversion, Vol.3, #1, January-June 2012, p. 43-69.
Book "Femtotechnology", Lulu, 2009. http://viXra.org/abs/1604.0271, http://www.archive.org/details/ConvertingOfAnyMatterToNuclearEnergyByAb-generatorAndAerospace , . http://intellectualarchive.com #1361, http://gsjournal.net/Science-Journals/Essays/View/5714 , http://GSJournal.net (10/3/14),. http://viXra.org/abs/1508.0305, http://viXra.org/abs/1604.0271, https://www.scribd.com/search-documents?query=Bolonkin

5) **Превращение материи в ядерную энергию АБ-генератором и фотонные ракеты** (in Russian) (популярное изложение научной статьи) .
http://www.archive.org/details/ConvertingOfMatterInMuclerEnergy. (see 26)
http://viXra.org/abs/1508.0307,

6) **RailGun Space Launcher,** Journal of Aerospace Engineering, Oct. 2010, Vol. 23, No. 4, pp.293-299.
http://viXra.org/abs/1401.0172, https://www.scribd.com/search-documents?query=Bolonkin , http://www.scribd.com/doc/202176675/Magnetic-Space-Launcher

7) **Artificial Explosion of Sun. AB-Criterion for Sun Detonation**.
Journal "Scientific Israel-Technological Advantages", Israel, Vol.13, #1, 2011, pp.45-64.
CWEEE, Computational Water, Energy, and Environmental Engineering. Volume 2, Number 3, July 2013 , Alexander Bolonkin, Joseph Friedlander Abstract | References PDF (372KB), PP. 83-96, Pub. Date: July 11, 2013 DOI: 10.4236/cweee.2013.23010, Open Access Library In Femto.
http://sita-journal.com/files/2_Cur.Iss_no.1.pdf , http://viXra.org/abs/1309.0198,
http://www.archive.org/details/ArtificialExplosionOfSun.Ab-criterionForSolarDetonation,
https://www.scribd.com/search-documents?query=Bolonkin. http://www.scribd.com/doc/24541542/

8) **Blanket for Cities.** Journal of Environmental Protection , Volume 02, Number 04 (June 2011). PP.327-341, Pub. Date:2011-06-17, http://viXra.org/abs/1309.0197, http://www.scribd.com/doc/24050198 http://www.archive.org/details/DomeShieldAMethodToContainRadioactiveDustFromDamagedNuclearStations , http://intellectualarchive.com #1363 .

9) **Man in Outer Space without a Special Space Suit**. (See 36)
American Journal of Engineering and Applied Science 2(4), 573-579, 2009, ISSN 1991-7020.
http://viXra.org/abs/1309.0199, https://www.scribd.com/search-documents?query=Bolonkin,
http://www.scribd.com/doc/24050793/,
http://www.archive.org/details/LiveOfHumanityInOuterSpaceWithoutSpaceSuite .
https://www.academia.edu/14515917/Man_in_Outer_Space_without_Space_Suite

10) **Aerial-High-Altitude-Gas-Pipeline**. The Open Petroleum Engineering Journal, 2009, 2, 24-35.
http://www.benthamscience.com/open/topej/articles/V002/24TOPEJ.pdf ,
http://www.archive.org/details/AerialAltitudeGasPipeline, http://intellectualarchive.com #1364 .
https://www.scribd.com/search-documents?query=Bolonkin, http://www.scribd.com/doc/24051138/ ,

11) **"Magnetic Space Launcher"** has been published online 15 December 2010, in the ASCE, Journal of Aerospace Engineering (Vol. 24, No. 1, 2011, pp.124-134).
http://www.scribd.com/doc/24051286/ , . https://www.scribd.com/search-documents?query=Bolonkin https://archive.org/details/MagneticSpaceLauncher (in PDF) 12) **An Innovative Solar Desalinization System**. http://www.scribd.com/doc/24051638/ Deleted .
Published in Badesky Collection together Neumann and Friedlander. Book "Femtotechnology",Lulu,2011.

13) **Economically Efficient Inflatable 3-km Tower for Communication**
http://www.scribd.com/doc/24051794/ Book "Femtotechnology", Lulu, 2011.

14) **Production-of-Fresh-Water-by-Exhaust-Gas-of-Electric-and-Industrial-Plants**.
Journal "Scientific Israel-Technological Advantages". Vol.13, #1, 2011, pp.65-71.
https://archive.org/details/ProductionOfFreshWaterByExhaustGas,
http://www.scribd.com/doc/24052023/ ,
http://intellectualarchive.com #1365 . Book "Femtotechnology", Lulu, 2011.

15) **Sea-Extractor-of-Freshwater**,
https://archive.org/details/ProductionOfFreshwaterAndEnergyFromAtmosphere ,
http://www.scribd.com/doc/24052153/

16) **AB-Wind-Wall**. Journal "Scientific Israel-Technological Advantages" , Vol.12, #4, 2010.
http://www.scribd.com/doc/24052321/ Book "Femtotechnology", Lulu, 2009.

17) **Utilization of Wind Energy at High Altitude.** Journal "Smart Grid and Renewable Energy", 2011,#2, pp.75-85. http://www.archive.org/details/UsingOfHighAltitudeWindEnergy ,
Book: New Concepts.

18) **Использование Энергии Ветра Больших Высот** (In Russian).
http://www.scribd.com/doc/24058357/

19) **"Magnetic Suspended AB-Structures and Motionless Space Stations**,"
has been published online 15 December 2010, in the ASCE, Journal of Aerospace Engineering (Vol.24,No.1, 2011, pp.102-111), https://archive.org/details/MagneticSuspendedAb-structures ,
http://www.scribd.com/doc/25883886/ . http://intellectualarchive.com #1366 .

20) **Природная цель Человечества – стать Богом**. http://www.scribd.com/doc/26753118 . (Russian). ?

21) **Natural Purpose of Mankind is to become a God**. http://www.scribd.com/doc/26833526
Book "Femtotechnology", by A.Bolonkin, Lulu, 2009. Ch.7B. ISBN: 978-1-300-448671-8.
http://viXra.org/abs/1309.0191,

22) **Magnetic Space AB-Accelerator**. http://www.scribd.com/doc/26885058 ,
Book: A.Bolonkin "Femtotechnology and Revolutionary Projects", Lambert, 2011.

23) **Lower Current and Plasma Magnetic Railguns.** http://www.scribd.com/doc/31090728
Propulsion: Types, Technology and Applications. NOVA. 2011. Book "Femtotechnology", Lulu,

2009. https://www.novapublishers.com/catalog/product_info.php?products_id=24848

24) **Review of Space Towers**. Book: A.Bolonkin, "Femtechnology and Revolutionary Projects", Lambert, 2011.. 538 p. 16 Mb.,Ch.13. ISBN: 978-3-8473-0839-0.
http://viXra.org/abs/1309.0191, http://arxiv.org/ftp/arxiv/papers/1002/1002.2405.pdf,
http://www.scribd.com/doc/75519828/, http://www.scribd.com/doc/26270139,
http://vixra.org/abs/1310.0009.

25) **Новые идеи в технологии, технике и оружии**. http://viXra.org/abs/1508.0308,
https://www.scribd.com/search-documents?query=Bolonkin. http://www.scribd.com/doc/27785947/,

26) **Превращение материи в ядерную энергию АБ-генератором и фотонные ракеты**
(популярное изложение научной статьи). http://www.scribd.com/doc/45901918/,
http://www.archive.org/details/ConvertingOfMatterInMuclerEnergy. (see 5)
http://viXra.org/abs/1508.0307,

27) **Wireless Transfer of Electricity from Continent to Continent**.
International Journal of Sustainable Engineering. 2011. Vol.4, #4, p. 290-300.
http://www.archive.org/details/WirelessTransferOfElectricityFromContinentToContinent
http://www.scribd.com/doc/42721638/, http://intellectualarchive.com #1367

28) **High-Altitude-Long-Distance-Cheap-Aerial-Antenna**.
http://www.scribd.com/doc/24052420/

29) **Robot as Person. Personhood. Three Prerequisites or Laws of Robots**. Book "Femtotechnology", Lulu, 2009. http://www.scribd.com/doc/57532296/Robot-as-Person-Personhood-Three-Prerequisites-or-Laws-of-Robots,
http://www.archive.org/details/RobotAsPersonpersonhood.ThreePrerequisitesOrLawsOfRobots

30) **Review of new ideas, innovations of non-rocket propulsion systems for Space Launch and Flight (Part 1)**. http://intellectualarchive.com #1368,
http://www.scribd.com/doc/54655572/,
http://www.archive.org/details/ReviewOfNewIdeasInnovationOfNon-rocketPropulsionSystemsForSpace
Book: A.Bolonkin,"Femtotechnology and Revolutionary Projects", Lambert, 2011. 538 pgs.

31) **Review of new ideas, innovations of non-rocket propulsion systems for Space Launch and Flight (Part 2)**. http://www.scribd.com/doc/54656166/, Book "Femtotechnology", Lulu, 2009.
http://www.archive.org/details/Review2OfNewIdeasInnovationsOfNonrocketPropulsionSystemsForSpace,
Book: A.Bolonkin "Femtotechnology and Revolutionary Projects", Lambert, 2011.

32) **Review of new ideas, innovations of non-rocket propulsion systems for Space Launch and Flight (Part 3)**. http://intellectualarchive.com #1369,
http://www.scribd.com/doc/54656800/,
http://www.archive.org/details/Review3OfNewIdeasInnovationsOfNon-rocketPropulsionSystemsForSpace,
Book "Femtotechnology", Lulu, 2009.
Book: A.Bolonkin "Femtotechnology and Revolutionary Projects", Lambert, 2011.

33) **Femtotechnology: AB-Needles. Fantastic properties and Applications.** 2010,
Journal of Energy Storige and Conversion, Vol.3, #1, January-June 2012, p. 15-41.
http://vixra.org/abs/1111.0064, http://intellectualarchive.com #1370,
http://www.archive.org/details/FemtotechnologyAb-needles.FantasticPropertiesAndApplications
http://vixra.org/pdf/1111.0064v1.pdf, http://www.scribd.com/doc/55054819/,,
Book: Femtotechnology, Lulu, 2009.
Published in collection: Propulsion: Types, Technology and Applications. NOVA. 2011.
https://www.novapublishers.com/catalog/product_info.php?products_id=24848

34) **АБ-материя и иглы.** Потрясающие свойства и применение (популярное изложение научной статьи). http://www.archive.org/details/Ab-matterAndAb-needles.FantasticProperties.-
http://www.pravda.ru/science/eureka/hypotheses/23-05-2011/1077601-nuclon_materia-0/,
http://viXra.org/abs/1508.0307

33) **Space Wing Electro Relativistic AB-Ship.**
I J N N A, 4(2) January-June 2012, pp. 13-19 • ISSN: 0974-6048, Collection "Femto"
Collection: Interstellar Medium: New Research. NOVA, 2011.
https://www.novapublishers.com/catalog/product_info.php?products_id=22357
http://www.scribd.com/doc/56874853/Space-Wing-Electro-Relativistic-AB-Ship

http://www.archive.org/details/SpaceWingElectroRelativisticAb-ship
http://intellectualarchive.com #1371,

34) **"Floating Cities on Ice Platform".** "The Open Ocean Engineering Journal". **Vol. 3, 2010, pp. 1-11.**
http://www.benthamscience.com/open/tooej/articles/V003/1TOOEJ.pdf ,
http://intellectualarchive.com #1371,

35) **Production of Freshwater and Energy from Earth's Atmosphere.** Journal "Smart Grid and Renewable Energy", 2011,#2, pp.86-98. http://www.scirp/jounal/sqgre/ . Book: New Consepts.2007
https://archive.org/details/ProductionOfFreshwaterAndEnergyFromAtmosphere_370 .
http://intellectualarchive.com #1372,

36) **Man in Outer Space without a Special Space Suit.** .(see 9)
American Journal of Engineering and Applied Sciences 2 (4): 573-579, 2009, ISSN 1941-7020.
http://www.scribd.com/doc/24050793 ,
https://www.nihms.nih.gov/db/sub.cgi?mid=688697
https://www.academia.edu/14515917/Man_in_Outer_Space_without_Space_Suite
Book: A.Bolonkin, "Femtotechnology and Revolutionary Projects", Lambert, 2011.??

37) **"Air Transfer of Mechanical Energy",** International Journal "Actual problems of aviation and aerospace systems: processes, models, experiments" (No. 1(19), v.10, 2005, pp.102-110). Coll. "Non-Rocket..."

38) **Suppression of Forest Fire by Helicopter without Water.** Journal "Scientific Israel- Technological Advantages" , Vol.12, 4, 2010. http://www.scribd.com/doc/24052503/,
http://viXra.org/abs/1410.0015
http://www.archive.org/details/SuppressionOfForestFireByHelicopterWithoutWater
Book: A.Bolonkin "Femtotechnology and Revolutionary Projects", Lambert, 2011.
http://intellectualarchive.com #1374,

39) **"Transparent Inflatable Column Film Dome for Nuclear Stations, Stadiums, and Cities,"** Science and Technology of Nuclear Installations, vol. 2011, Article ID 175492, 13 pages, 2011. doi:10.1155/2011/175492. http://www.hindawi.com/journals/stni/2011/175492/

40) **Problems of Science Research and Technical Progress.** 2011, 15 pages.
Together with Neumann and Fridlander. Scientific Israel - Technological
Advantages Vol. 14, No 1, 2012. http://www.scribd.com/doc/74436485/,
http://vixra.org/pdf/1112.0001v1.pdf ,

41) **Air Catapult Transportation**. NY, USA, Archuve, 2011 IJEE.. http://viXra.org/abs/1310.0065 ,
http://www.archive.org/details/AirCatapultTransport, http://intellectualarchive.com #1375,
http://www.scribd.com/doc/79396121/Article-Air-Catapult-Transportation-for-Scribd-1-25-12,
Book: Recent Patents on Electrical & Electronic Engineering, Bentham Science Publishers, Vol.5, No.3, 2012.

42) **Long Distance Bullets and Shells.**
International Journal of Aerospace Sciences. p-ISSN: 2169-8872, e-ISSN:2169-8899. 2013; 2(2): 29-36, http://viXra.org/abs/1207.0012 (?)
http://archive.org/details/LongDistanceBulletsAndShells , http://intellectualarchive.com #1375,
http://www.scribd.com/doc/99132995/Long-Distance-Bullets-and-Shells ,

43) **New Self-Propelled Penetration Bomb.** International Journal of Advanced Engineering Applications, Vol.2, Iss.5, pp.91-105 (2013).
http://fragrancejournals.com/wp-content/uploads/2013/03/IJAEA-2-5-14.pdf
http://www.scribd.com/doc/99131896/NEW-SELF-PROPELLED-PENETRATION-BOMB
http://archive.org/details/NewSelf-propelledPenetrationBomb, http://intellectualarchive.com #137, http://viXra.org/abs/1207.0013|

44) **Delivery of Asteroids to the Earth**. IJES, Vol.3, No.2, July-December 2012, pp.55-62.
http://archive.org/details/CaptureAndDeliveryOfAsteroidToTheEarth .
https://archive.org/details/TransportationOfAsteroidToTheEarth, http://intellectualarchive.com #138,

http://www.scribd.com/doc/99132263/Capture-and-Delivery-of-Asteroid-to-the-Earth,
http://viXra.org/abs/1207.0011

45) **Universe (part 1). Relations between Time, Matter, Volume, Distance, and Energy.**
JOURNAL OF ENERGY STORAGE AND CONVERSION, JESC : July-December 2012, Volume 3, #2, pp. 141-154. http://viXra.org/abs/1207.0075 , http://intellectualarchive.com #139, http://www.scribd.com/doc/100541327/ ,
http://archive.org/details/Universe.RelationsBetweenTimeMatterVolumeDistanceAndEnergy

46) **Universe (Part 2): Rolling of Space (Volume, Distance), Time, and Matter into a Point.**
http://www.scribd.com/doc/120693979 . No.
https://archive.org/download/Universepart2RollingOfSpacevolumeDistanceTimeAndMatterIntoA

47) **"Remarks about Universe" (part 1-2)**, International Journal of Advanced Engineering Applications, IJAEA. Vol.1, Iss.3, pp.62-67 (2012) , http://viXra.org/abs/1309.0196 , OK 8 5 17, http://fragrancejournals.com/wp-content/uploads/2013/03/IJAEA-1-3-10.pdf ???

48) **Cheap Protection of New York City and New Jersey from Storm.**
https://archive.org/details/CheapProtectionOfCityAndOtherPlaceFromFloodAndHurricane,
http://intellectualarchive.com #1380 .
http://www.scribd.com/doc/112009139/Article-Protection-of-NY-From-Storm-2006 (In Macro)

49) **Protection of the Earth from Asteroids.** http://intellectualarchive.com #1381,
http://viXra.org/abs/1212.0006,
http://archive.org/details/ProtectionOfTheEarthFromTheAsteroid,
http://www.scribd.com/doc/115171595/Protection-of-the-Earth-from-the-Asteroid

50) **Re-Entry Space Apparatus to Earth.** General Science Journal, #5289.
http://www.gsjournal.net/Science-Journals/Research%20Papers-Engineering%20(Applied)/Download/5289
http://archive.org/details/ReentryOfSpaceCraftToEarthAtmosphere,
http://intellectualarchive.com #1381,
http://www.scribd.com/doc/115174092/REENTRY-OF-SPACE-CRAFT-TO-EARTH-ATMOSPHERE, http://viXra.org/abs/1212.0003

51) **Hypersonic Ground Electric AB Engine.**
Academic Journal of Applied Sciences Research (AJASR), Volume-1, Issue–1, 2013.
http://archive.org/details/HypersonicGroundElectricAbEngine,
http://www.scribd.com/doc/119462908/Hypersonic-Ground-Electric-AB-Engine,
http://intellectualarchive.com #1383,
International Journal of Advanced Engineering Applications, Vol.1, Iss.4, pp.32-43 (2012)
http://fragrancejournals.com/wp-content/uploads/2013/03/IJAEA-1-4-4.pdf,
https://archive.org/details/HypersonicCatapultTransportation .

52) **"Production of Freshwater and Energy from Earth's Atmosphere"**, Smart Grid and Renewable Energy (SGRE), Vol.02 No.02, 2011. 826 downloads and the citations based on the statistics from Google Scholar, Please access the following link: Google Scholar (2011). Col. "New Consepts…"
https://archive.org/details/ProductionOfFreshwaterAndEnergyFromAtmosphere_370 .
http://intellectualarchive.com #1384,

53) **"Protection of Environment from Damaged Nuclear Station and Transparent Inflatable Blanket**
for Cities. Protection from Radioactive Dust and Chemical, Biological Weapons", Journal of Environmental Protection (JEP), Vol. 02 No.04, 2011 , Article has 738 downloads and the

citations based on the statistics from Google Scholar, Please access the following link: Google Scholar (2011).

54) **Energy Transfers from Airborne Wind Turbine: Review and Comparison of Airborne Turbines.** International Journal of Advanced Engineering Applications, 2013, Vol.1, Iss.4, pp.44-64. http://viXra.org/abs/1304.0159,
http://archive.org/details/EnergyTransfersFromAirborneWindTurbine,
http://www.scribd.com/doc/138350864/Energy-Transfers-from-Airborne-Wind-Turbine-Review-and- Comparison-of-Airborne-Turbines-Article-Transfer-energy-from-air-borne-turb ,
http://fragrancejournals.com/wp-content/uploads/2013/03/IJAEA-1-4-5.pdf

55) **Underground Explosion Nuclear Energy**.
International Journal of Advanced Engineering Applications, Vol.1, Iss.6, pp.48-61 (2012).
www.IntellectualArchive.com/getfile.php?file=TOe6vifJr1D&orig_file=Article Explosion Nuclear Energy2 for Storage 3 8 13.doc, http://viXra.org/abs/1305.0039 ,
http://archive.org/details/UndergroundExplosionNuclearEnergy, http://intellectualarchive.com #1385, http://fragrancejournals.com/wp-content/uploads/2013/03/IJAEA-1-6-7.pdf ,
http://www.scribd.com , www.fragrancejournals.com/IJAEA, Request #284619 .

56) **Inexpensive Mini Thermonuclear Reactor.**
International Journal of Advanced Engineering Applications, Vol.1, Iss.6, pp.62-77 (2012)
http://archive.org/details/InexpensiveMiniThermonuclearReactor, http://viXra.org/abs/1305.0046
www.IntellectualArchive.com/getfile.php?file=gIhLJg6ZAaN&orig_file=Article Thermonuclear Reactor for Storage 5 7 13.doc   (They made "Privet")
http://ru.scribd.com/doc/140040026/Inexpensive-Mini-Thermonuclear-Reactor-Article-Thermonuclear-Reactor-for-Storage-5-7-13, http://fragrancejournals.com/wp-content/uploads/2013/03/IJAEA-1-6-8.pdf

57) **Electron Air Hypersonic Propulsion**. International Journal of Advanced Engineering Applications, Vol.1, Iss.6, pp.42-47 (2012). http://viXra.org/abs/1306.0003,
http://intellectualarchive.com #1386,
http://www.scribd.com/doc/145165015/Electron-Air-Hypersonic-Propulsion ,
http://www.scribd.com/doc/146179116/Electronic-Air-Hypersonic-Propulsion ,
http://fragrancejournals.com/wp-content/uploads/2013/03/IJAEA-1-6-6.pdf .

58) **Electronic Wind Generator**.
Electrical and Power Engineering Frontier Sep. 2013, Vol. 2 Iss. 3, pp. 64-71.
http://www.academicpub.org/epef/Issue.aspx?Volume=2&Number=3&Abstr=false
http://viXra.org/abs/1306.0046 , www.IntellectualArchive.com, http://intellectualarchive.com #1387,
https://archive.org/details/ArticleElectronWindGenerator6613AsterShmuelWithPicture ,
http://www.scribd.com/doc/146177073/Electronic-Wind-Generator ?

59) **Electron Hydro Electric Generator.** International Journal of Advanced Engineering Applications. ISSN: 2321-7723 (Online), Special Issue I, 2013.
http://fragrancejournals.com/?page_id=18, http://viXra.org/abs/1306.0196,
http://www.scribd.com/doc/149489902/Electron-Hydro-Electric-Generator , #1089
http://archive.org/details/ElectronHydroElectricGenerator_532, http://intellectualarchive.com,

60) **Electron Super Speed Hydro Propulsion.**
International Journal of Advanced Engineering Applications, Special Issue 1, pp.15-19 (2013)
http://viXra.org/abs/1306.0195, http://archive.org/details/ElectronSuperSpeedHydroPropulsion
http://www.scribd.com/doc/149490731/Electron-Super-Speed-Hydro-Propulsion
http://intellectualarchive.com, Search: Bolonkin #1090

http://fragrancejournals.com/wp-content/uploads/2013/03/Special-Issue-1-4.pdf

61) **Electric Theory of Tornado. Protection from Tornado**.
International Journal of Advanced Engineering Applications. ISSN: 2321-7723 (Online), Volume 2, Issue 5 (October, 2013). http://fragrancejournals.com/?page_id=18 ,
http://www.scribd.com/doc/153430778/Electric-Theory-of-Tornado-Protection-from-Tornado
https://archive.org/details/ArticleElectricTheoryOfTornado2ForStorige7913
http://viXra.org/abs/1307.0061 , http://intellectualarchive.com/?link=find#detail #1100.
https://www.academia.edu/s/87c3793ac7

62) **Femtotechnology. AB-matter. Properties, Stability, Possibility Production and Applications.** EPEF, GBS, Con.FEMR, RAEEE (answer in Info 1 28 14)

63) **Stability and Production Super-Strong AB Matter**. International Journal of Advanced Engineering Applications. 3-1-3, February 2014, pp.18-33.
http://fragrancejournals.com/wp-content/uploads/2013/03/IJAEA-3-1-3.pdf
The General Science Journal, November, 2013, #5244.
http://www.gsjournal.net/Science-Journals/Research%20Papers-Quantum%20Theory%20/%20Particle%20Physics/Download/5244
http://www.scribd.com/doc/193675800/Stability-and-Production-Super-Strong-AB-matter ,
http://viXra.org/abs/1312.0017. https://archive.org/details/StabilityAndProductionSuper-strongAb-matter ,
http://www.IntellectualArchive.com/ Reference #1178,
GSJ 1 8 14. https://www.academia.edu/14514987/Stability_and_Production_Super-Strong_AB-matter

64) **Universe (Part 3). Relations between Charge, Time, Matter, Volume, Distance, and Energy.**
The General Science Journal, #5245. IJAEA, GSJ,
http://www.gsjournal.net/Science-Journals/Research%20Papers-Mechanics%20/%20Electrodynamics/Download/5245 , http://viXra.org/abs/1401.0075,
http://www.scribd.com/doc/197830994, http://www.IntellectualArchive.com/ Reference #1192 ,
https://archive.org/details/universepart3.RelationsBetweenChargeTimeMatterVolumeDistance
http://www.IntellectualArchive.com/getfile.php?file=gwBJfgtbOeS&orig_file=Article Universe3 after Friedlander 01 09 14.doc , https://www.academia.edu/14514621/Universe_Part_3_._Relations_between_Charge

65) **Provisional patent application "Method and installation for cleaning the outer debris"**
http://vixra.org/abs/1403.0669 , http://intellectualarchive.com/?link=find#result #1242,
http://www.scibd/doc/213887553 , http://www.IntellectualArchive.com #1388,

66) **Electric Hypersonic Space Aircraft.** http://intellectualarchive.com, #1288;
http://vixra.org/abs/1407.0011 , 1 July 2014; http://www.scibd/doc/232209230,
http://archive.org/details/ElectricHypersonicaircraft, http://gsjournal.net/Science-Journals-Papers/Author/1481/Alexander,%20Bolonkin .

67) **Electrostatic Generator and Electric Transfomer,** http://Vixra.org/abs/1407.0016,
http://GSJournal.net, 2 July 2014;   http://intellectualarchive.com, #1289;

68) **Jet generator.** http://Vixra.org/abs/1407.0180 , https://archive.org/details/,
http://gsjournal.net/Science-Journals-Papers/Author/1481/Alexander,%20Bolonkin

69) **Method for Interstellar Flight.** http://Vixra.org/abs/1408.0055, https://archive.org/details/
, http://intellectualarchive.com, #1312; http://gsjournal.net/Science-Journals-Papers/Author/1481/Alexander,%20Bolonkin

70) **Terroformating of planets and Space Objects.** http://gsjournal.net/Science-Journals-Papers/Author/1481/Alexander,%20Bolonkin , http://Vixra.org/abs/1408.0239,
http://archive.org/details/TerroformatingOfPlanetsAndSpaceObjects , http://intellectualarchive.com,
#1323.

71) **Cumulative Thermonuclear AB-Reactor.** Vixra 7 8 15, http://viXra.org/abs/1507.0053
https://archive.org/details/ArticleCumulativeReactorFinalAfterCathAndOlga7716 ,
http://intellectualarchive.com, #1547, GSJ 7 9 15, GSJornal: http://gsjournal.net/Science-Journals/%7B$cat_name%7D/View/6134,

https://www.researchgate.net/profile/Alexander_Bolonkin/publications?sorting=recentlyAdded
www.IntellectualArchive.com/getfile.php?file=QDvULGMdCBU&orig_file=Article Cumulative Reactor Final after Cath and Olga 3 7 16.docx ,
https://www.academia.edu/14510693/Cumulative_Thermonuclear_AB-Reactor

Journal: *Energy, Sustamobility and Society,* Springer, v.6, issue 1, 2016. DOI: 10.1186/s13705-016-0074-z, ESSO-D-15-00052.1

72) **Ultra-Cold Thermonuclear Synthesis: Criterion of Cold Fusion.** 7 18 15. 5/5/16 sibmitted in J. "Engineering" Refusel – publised. http://viXra.org/abs/1507.0158 , https://archive.org/details/ArticleColdFusionAfterRichard71815; www.IntellectualArchive.com , #1556; GSJornal: http://gsjournal.net/Science-Journals/%7B$cat_name%7D/View/6140 , https://www.academia.edu , https://www.researchgate.net/profile/Alexander_Bolonkin

73) **Impulse solutions in optimization problems** 11 20 15

http://viXra.org/abs/1511.0189; https://www.academia.edu/s/3244c0c4f0?source=link ???
https://www.academia.edu/s/00538971c8 ,
https://archive.org/details/ArticleImpulseSolutionsdoc200311115AfterJoseph
GSJornal: http://gsjournal.net/Science-Journals/Research%20Papers-Astrophysics/Download/6259
www.IntellectualArchive.com, #1625, https://www.researchgate.net ???

74) **AB Preon Interaction Theory and Model of Universe,** http://vixra.org/abs/1603.0210, https://www.academia.edu , 3 30 16 Journal GSJ, http://gsjournal.net
https://archive.org/details/ArticlePreonUniverseAfterJoseph22516, www.IntellectualArchive.com #1698 5/5/16, I submeeted to J. "Engineering" (refusal: published in GSJ). Lulu published the Collection on March 2017

75) **Cumulative and Impulse Mini Thermonuclear Reactors.** 3 30 16, http://viXra.org/abs/1605.0309 , 5 31 16.
https://archive.org/download/ImpulseMiniThermonuclearReactors , 5 31 16.

76) **Electric Cumulative Thermonuclear Reactors.** 7 17 16. 10 17 16.
http://vixra.org/abs/1610.0208, https://archive.org/download/abolonkin_gmail_201610 ,

www.IntellectualArchive.com , Bolonkin, 1771 ,

77) **Tritium Fusion Energy is the Costliest Mistake in the History of Science.** 9 11 16

Sent to: Engineering, New Space, Physics CogentOA 9 21 16. Engenering, New Space –refusel.

Submitted: Fusion Engineering and Design, Codent Physics.-refusel. www.Vixra.org 2 15 17, #9246940, https://archive.org/details/ArticleTritiumFusionMistakeAfterZarek916161

78) **Small, Non-Expensive Electric Impulse Thermonuclear Reactor with colliding jets.** 7 11 16, 11 19 16, http://viXra.org/abs/1611.0276 ,
https://archive.org/download/ArticleThermonuclearReactorOfCollisingJets10416,

IGJournal 3 2 17, www.IntellectualArchive.com #1817, 3 2 17. Lulu published in March 2017

79) **Flight of Outer Solar System**. 5 2 16.
https://archive.org/download/ChapterFlightOfEnergyInOutSolarSystem111516

80) **Sources of Energy in the Outer Solar System,** 6 1 16.
https://archive.org/download/ChapterSoursesOfEnergyInOutSolarSystem111516

81) **Transparent Fuel Capsule for Fusion Reactor** 2 14 17. http://viXra.org/abs/1702.0170
https://archive.org/details/ArticleTransparentFuelCapsuleAfterSpellingAndRichard2131712

www.IntellectualArchive.com , #1807 2 14 17. IGJ 3 2 17,
www.IntellectualArchive.com #1816

82) **Universe (Part 4). Relations between Charge, Time, Matter, Volume, Distance, and Energy.** http://viXra.org/abs/1708.0062 , www.IntellectualArchive.com #1860,
https://archive.org/details/ArticleUniverse4062217 , GIJ.

Some Collections having Bolonkin's articles:

0) Editor Badescu "**Resurces Outer Solar System**". Elsevier.2017
   1. Flight of Outer Solar System
   2. Sources of Energy in the Outer Solar System, 6 1 16

1) **Macro-engineering Seawater in Unique Environments. Springer. 2010.**
http://www.springer.com/environment/aquatic+sciences/book/978-3-642-14778-4

   1. The Bering Strait Seawater Deflector (BSSD): Arctic Tundra Preservation Using an Immersed, Scalable and Removable Fiberglass Curtain. 741
   Richard B. Cathcart, Alexander A. Bolonkin and Radu D. Rugescu

   2. A Novel Macro-Engineering Approach to Seawater Desalination 675
   Alexander A. Bolonkin, Shmuel Neumann and Joseph J. Friedlander

   3. **Macro-Engineering Lake Eyre with Imported Seawater** 553
   Viorel Badescu, Richard B. Cathcart, Marius Paulescu, Paul Gravila and Alexander A. Bolonkin

2) **Handbook on Solar Wind: Effects, Dynamics and Interactions. NOVA. 2009.**
https://www.novapublishers.com/catalog/product_info.php?products_id=8903
   1. Electrostatic Solar Light - Wind Sail, pp. 353-365 (Alexander Bolonkin, C & R, Brooklyn, NY)
   2. AB - Solar and Solar Wind Sail, pp. 367-378. (Alexander Bolonkin).
   3. Electrostatic Magsail.p. 379-389 . (Alexander Bolonkin,).

3) **Interstellar Medium: New Research. NOVA, 2011.**

https://www.novapublishers.com/catalog/product_info.php?products_id=22357
   1) Space Wing Electro AB-Ship. (Alexander Bolonkin, C&R, Brooklyn, New York, USA).

4) **Propulsion: Types, Technology and Applications. NOVA. 2011.**
https://www.novapublishers.com/catalog/product_info.php?products_id=24848
   1) Review of New Ideas, Innovation of Non-Rocket Propulsion Systems for Space Launch and Flight - (Part 1) . (Alexander Bolonkin, C&R, Brooklyn, New York, USA)
   2) Review of New Ideas, Innovations of Non-Rocket Propulsion Systems for Space Launch and Flight

- (Part 2). (Alexander Bolonkin, C&R, NJIT, Brooklyn, New York, USA)
3) Review of New Ideas, Innovations of Non-Rocket Propulsion Systems for Space Launch and Flight – (Part 3). (Alexander Bolonkin, C&R, NJIT, Brooklyn, New York, USA)
4) Superconductivity Space Accelerator. (Alexander Bolonkin, C&R, Brooklyn, New York, USA)
5) Femtotechnology: AB-Needles - Fantastic Properties and Applications in Propulsion System and Aerospace. (A.A. Bolonkin, C&R, Brooklyn New York, USA)
6) Lower Current and Plasma Magnetic RailGun. (A. Bolonkin, C&R, Brooklyn, New York, USA)

5) Collection: **Mars**. Springer. 2009.
  Ch.10. New Solutions for Nuclear Energy and Flights on Mars, pp.287-330.
  Ch.23. Artificial Environments on Mars. Pp.599-628.

6) Collection: **Macro-Engineering**. A Challenge for the Future. Springer. 2006.
  1) Space Towers. Pp.121-150.
  2) Cable Anti-Gravitator. Electrostatic Levitation and Artificial Gravity. Pp.175-214.

7) Collection: **Asteroids**. Perspective, Energy, and Material Resources, Springer, 2012. ISBN 078-3-642-39244-3
  1) Change the Asteroid rajectory.
  2) Shpad Metal Earth –Delivery Systems.
  3) Artificial Gravitation on Asteroids.
  4) Making Asteroids Habitable.
  5) Using Asteroids for Launch/Landing. Change of Trajectory and Acceleration of Space Ships.

7) Collection: **Inner Solar System.** Prospective Energy and Material Resources. Springer. 2015. ISBN 978-3-319-19569-8 (eBook).

  1) Estimation of the Fuel Consumption for Space Trip to Mercury and Venus.
  2) Production of Energy for Venus by Electron Wind Generator.
  3) Flight Apparatuses and Balloons in Venus Atmosphere.
  4) Artificial Magnetic Field for Venus.
  5) Ecomomic Development of Mercury: A Comparison with Mars Colonization.
  6) Terraforming Mercury and Venus.

8) Collection: **Asteroids**. Prospective Energy and Material Resources. Springer. 2014.
  ISBN 978-3-642-39244-3 (eBook).

  1) Change the Asteroid Trajectory.
  2) Shaped Metal Earth-Delivery Systems.
  3) Artificial Gravitation on Asteroids.
  4) Making Asteroids Habitable.
  5) Usinf Asteroids for Launch/Landing, Change of Trajectory and Acceleration of Space Ships.

There are a lot of articles published in Journals: IBIS, GSJ. Arxiv, Vixra, Archive, Academia,onferences AIAA, World Space Congress, etc.